Mike McGrath

C++
Programming

Sixth Edition

In easy steps is an imprint of In Easy Steps Limited
16 Hamilton Terrace · Holly Walk · Leamington Spa
Warwickshire · United Kingdom · CV32 4LY
www.ineasysteps.com

Sixth Edition

In Easy Steps Limited supports The Forest Stewardship Council (FSC),
the leading international forest certification organization. All our titles
that are printed on Greenpeace approved FSC certified paper carry the
FSC logo.

MIX
Paper from
responsible sources
FSC® C020837

FSC
www.fsc.org

Printed and bound in the United Kingdom

ISBN 978-1-84078-971-3

Contents

Preface

The creation of this book has provided me, Mike McGrath, a welcome opportunity to update my previous books on C++ programming with the latest techniques. All examples I have given in this book demonstrate C++ features supported by current compilers on both Windows and Linux operating systems, and in the Microsoft Visual Studio development suite. The book's screenshots illustrate the actual results produced by compiling and executing the listed code. I sincerely hope you enjoy discovering the powerful, expressive possibilities of C++ Programming and have as much fun with it as I did in writing this book.

Conventions in this book

In order to clarify the code listed in the steps given in each example, I have adopted certain colorization conventions. Components of the C++ language itself are colored blue, numeric and string values are red, programmer-specified names are black, and comments are green, like this:

```cpp
// Store then output a text string value.
string myMessage = "Hello from C++!" ;
cout << myMessage ;
```

Additionally, in order to identify each source code file described in the steps, a colored icon and file name appears in the margin alongside the steps:

main.cpp   header.h

Grabbing the source code

For convenience I have placed source code files from the examples featured in this book into a single ZIP archive, providing versions for Windows and Linux platforms plus the Microsoft Visual Studio IDE. You can obtain the complete archive by following these easy steps:

1. Browse to **www.ineasysteps.com** then navigate to Free Resources and choose the Downloads section

2. Find C++ Programming in easy steps, 6th edition in the list then click on the hyperlink entitled All Code Examples to download the archive

3. Now, extract the archive contents to any convenient location on your computer

If you don't achieve the result illustrated in any example, simply compare your code to that in the original example files you have downloaded to discover where you went wrong.

1 Getting started

Welcome to the exciting world of C++ programming. This chapter demonstrates how to create a simple C++ program and how to store data within a program.

Introducing C++

A powerful programming language (pronounced "see plus plus"), designed to let you express ideas.

C++ is an extension of the C programming language that was first implemented on the UNIX operating system by Dennis Ritchie way back in 1972. C is a flexible programming language that remains popular today, and is used on a large number of platforms for everything from microcontrollers to the most advanced scientific systems.

C++ was developed by Dr. Bjarne Stroustrup between 1983 and 1985 while working at AT&T Bell Labs in New Jersey. He added features to the original C language to produce what he called "C with classes". These classes define programming objects with specific features that transform the procedural nature of C into the object-oriented programming language of C++.

The C programming language was so named as it succeeded an earlier programming language named "B" that had been introduced around 1970. The name "C++" displays some programmers' humor because the programming ++ increment operator denotes that C++ is an extension of the C language.

C++, like C, is not platform-dependent, so programs can be created on any operating system. Most illustrations in this book depict output on the Windows operating system purely because it is the most widely used desktop platform. The examples can also be created on other platforms such as Linux or macOS.

Why learn C++ programming?

The C++ language is favored by many professional programmers because it allows them to create fast, compact programs that are robust and portable.

Microsoft's free Visual Studio Community Edition IDE is used in this book to demonstrate visual programming.

Using a modern C++ Integrated Development Environment (IDE), such as Microsoft's Visual Studio Community Edition, the programmer can quickly create complex applications. But to use these tools to greatest effect, the programmer must first learn quite a bit about the C++ language itself.

This book is an introduction to programming with C++, giving examples of program code and its output to demonstrate the basics of this powerful language.

Should I learn C first?

Opinion is divided on the question of whether it is an advantage to be familiar with C programming before moving on to C++. It would seem logical to learn the original language first in order to understand the larger extended language more readily. However, C++ is not simply a larger version of C, as the approach to object-oriented programming with C++ is markedly different to the procedural nature of C. It is, therefore, arguably better to learn C++ without previous knowledge of C to avoid confusion.

This book makes no assumption that the reader has previous knowledge of any programming language, so it is suitable for the beginner to programming in C++, whether they know C or not.

If you do feel that you would benefit from learning to program in C before moving on to C++, we recommend you try the examples in **C Programming in easy steps** before reading this book.

Standardization of C++

As the C++ programming language gained in popularity, it was adopted by many programmers around the world as their programming language of choice. Some of these programmers began to add their own extensions to the language, so it became necessary to agree upon a precise version of C++ that could be commonly shared internationally by all programmers.

A standard version of C++ was defined by a joint committee of the American National Standards Institute (ANSI) and the Industry Organization for Standardization (ISO). This version is sometimes known as ANSI C++, and is portable to any platform and to any development environment.

The examples given in this book conform to ANSI C++. Example programs run in a console window, such as the Command Prompt window on Windows systems or a shell terminal window on Linux systems, to demonstrate the mechanics of the C++ language itself. An example in the final chapter illustrates how code generated automatically by a visual development tool on the Windows platform can, once you're familiar with the C++ language, be edited to create a graphical, windowed application.

Hot tip

"ISO" is not an acronym but is derived from the Greek word "isos" meaning "equal" – as in "isometric".

Installing a compiler

C++ programs are initially created as plain text files, saved with the file extension of ".cpp". These can be written in any text editor, such as Windows' Notepad application or the Vi editor on Linux.

In order to execute a C++ program, it must first be "compiled" into byte code that can be understood by the computer. A C++ compiler reads the text version of the program and translates it into a second file – in machine-readable, executable format.

Should the text program contain any syntax errors, these will be reported by the compiler and the executable file will not be built.

If you are using the Windows platform and have a C++ Integrated Development Environment (IDE) installed, then you will already have a C++ compiler available, as the compiler is an integral part of the visual IDE. The excellent, free Microsoft Visual C++ Express IDE provides an editor window, where the program code can be written, and buttons to compile and execute the program. Visual IDEs can, however, seem unwieldy when starting out with C++ because they always create a large number of "project" files that are used by advanced programs.

The popular free GNU C++ Compiler is included with most distributions of the Linux operating system. The GNU C++ Compiler is also available for Windows platforms and is used to compile examples throughout this book.

To discover if you already have the GNU C++ Compiler on your system, type **c++ -v** at a command prompt then hit **Return**. If it's available, the compiler will respond with version information. If you are using the Linux platform and the GNU C++ Compiler is not available on your computer, install it from the distribution disc, download it from the GNU website, or ask your system administrator to install it.

The GNU (pronounced "guh-new") Project was launched back in 1984 to develop a complete free Unix-like operating system. Part of GNU is "Minimalist GNU for Windows" (MinGW). MinGW includes the GNU C++ Compiler that can be used on Windows systems to create executable C++ programs. Windows users can download and install the GNU C++ Compiler by following the instructions on the opposite page.

The GNU C++ compiler is available free under the terms and conditions of the General Public License (GPL) that can be found online at **gnu.org/copyleft/gpl.html**

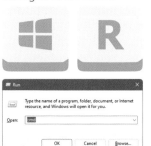

To open a Windows Command Prompt, press the **Windows** + **R** keys to launch a Run dialog, then type **cmd** into the dialog and hit **Enter**.

1 With an internet connection, launch a web browser then navigate to **osdn.net/projects/mingw** and click the link to download the MinGW installer **mingw-get-setup.exe**

2 Launch the installer setup and accept the suggested location of **C:\MinGW** in the "Installation Manager" dialog

3 Check the **Basic MinGW** and **C++ Compiler** items, then click **Installation**, **Apply Changes**, **Apply** to install

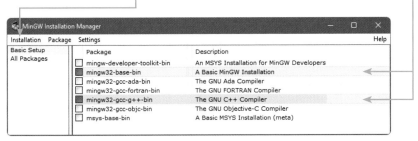

The MinGW C++ Compiler is a binary executable file located at **C:\MinGW\bin**. To allow it to be accessible from any system location, this folder should now be added to the System Path:

4 Open Windows' "System Properties" dialog, then select the **Advanced** tab and click the **Environment Variables** button – to open the "Environment Variables" dialog

5 Select the **Path** system variable, then click the **Edit** button and add the location **C:\MinGW\bin;**

6 Click **OK** to close each dialog, then open a Command Prompt window and enter the command **c++**. If the installation is successful, the compiler should respond that you have not specified any input files for compilation:

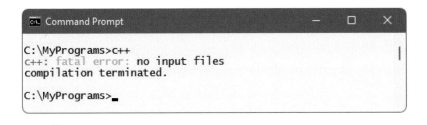

Hot tip

To open a System Properties dialog, press the **Windows** + **R** keys to launch a Run dialog, then type **sysdm.cpl** into the dialog and hit **Enter**.

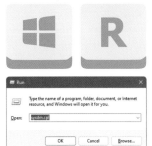

Writing your first program

Follow these steps, copying the code exactly as it is listed, to create a simple C++ program that will output the traditional first program greeting:

hello.cpp

Comments throughout this book are shown in green – to differentiate them from other code.

1   Open a plain text editor, such as Windows' Notepad, then type these "preprocessor directives"
```
#include <iostream>
using namespace std ;
```

2   A few lines below the preprocessor directives, add a "comment" describing the program
```
// A C++ Program to output a greeting.
```

3   Below the comment, add a "main function" declaration to contain the program statements
```
int main()
{

}
```

4   Between the curly brackets (braces) of the main function, insert this output "statement"
```
cout << "Hello World!" << endl ;
```

5   Next, insert a final "return" statement in the main function
```
return 0 ;
```

6   Save the program to any convenient location as "hello.cpp" – the complete program should look like this:

Beware

After typing the final closing **}** brace of the main method, always hit **Return** to add a newline character – your compiler may insist that a source file should end with a newline character.

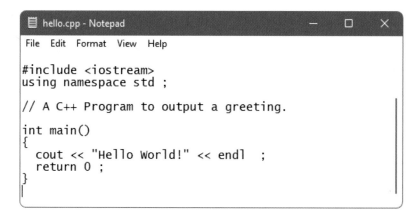

```
hello.cpp - Notepad
File  Edit  Format  View  Help

#include <iostream>
using namespace std ;

// A C++ Program to output a greeting.

int main()
{
  cout << "Hello World!" << endl  ;
  return 0 ;
}
```

The separate parts of the program code on the opposite page can be examined individually to understand each part more clearly:

- **Preprocessor Directives** – these are processed by the compiler before the program code, so must always appear at the start of the page. Here, the **#include** directive instructs the compiler to use the standard C++ input/output library named **iostream**, specifying the library name between **< >** angled brackets. The next line is the "using directive" that allows functions in the specified namespace to be used without their namespace prefix. Functions of the **iostream** library are within the **std** namespace – so this **using** directive allows functions such as **std::cout** and **std::endl** to be simply written as **cout** and **endl**.

- **Comments** – these should be used to make the code more easily understood by others, and by yourself when revisiting the code later. In C++ programming, everything on a single line after a **//** double-slash is ignored by the compiler.

- **Main function** – this is the mandatory entry point of every C++ program. Programs may contain many functions, but they must always contain one named **main**, otherwise the compiler will not compile the program. Optionally, the parentheses after the function name may specify a comma-separated list of "argument" values to be used by that function. Following execution, the function must return a value to the operating system of the data type specified in its declaration – in this case, an **int** (integer) value.

- **Statements** – these are the actions that the program will execute when it runs. Each statement must be terminated by a semi-colon, in the same way that English language sentences must be terminated by a period (full stop). Here, the first statement calls upon the **cout** library function to output text and an **endl** carriage return. These are directed to standard output by the **<<** output stream operator. Notice that text strings in C++ must always be enclosed within double quotes. The final statement employs the C++ **return** keyword to return a zero integer value to the operating system – as required by the main function declaration. Traditionally, returning a zero value indicates that the program executed successfully.

Hot tip

The C++ compiler also supports multiple-line C-style comments between /* and */ – but these should only ever be used in C++ programming to "comment-out" sections of code when debugging.

Hot tip

Notice how the program code is formatted using spacing and indentation (collectively known as whitespace) to improve readability. All whitespace is ignored by the C++ compiler.

Compiling & running programs

The C++ source code files for the examples in this book are stored in a directory created expressly for that purpose. The directory is named "MyPrograms" – its absolute address on a Windows system is **C:\MyPrograms** and on Linux it's **/home/*user*/MyPrograms**. You can recreate this directory to store programs awaiting compilation:

You can see the compiler version number with the command
c++ --version and display all its options with **c++ --help**

1   Move the "hello.cpp" program source code file, created on page 12, to the "MyPrograms" directory on your system

2   At a command prompt, use the "cd" command to navigate to the "MyPrograms" directory

3   Enter a command to attempt to compile the program
c++ hello.cpp

When the attempt succeeds, the compiler creates an executable file alongside the original source code file. By default, the executable file is named **a.exe** on Windows systems and **a.out** on Linux. Compiling a different source code file in the same directory would now overwrite the first executable file without warning. This is obviously undesirable, so a custom name for the executable file should be specified when compiling programs, using the compiler's **-o** option in the compile command.

4   Enter a command to compile the program, creating an executable file named "hello.exe" alongside the source file
c++ hello.cpp -o hello.exe

The command **c++** is an alias for the GNU C++ Compiler – the command **g++** can also be used.

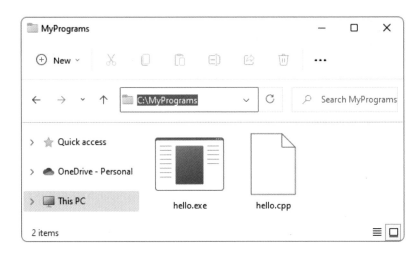

14

5   To run the generated executable program file in Windows, simply enter the file name at the prompt in the "MyPrograms" directory – optionally, the file extension may be omitted. In Linux, the full file name must be used, preceded by a ./ dot-slash – as Linux does not look in the current directory unless it is explicitly directed to do so:

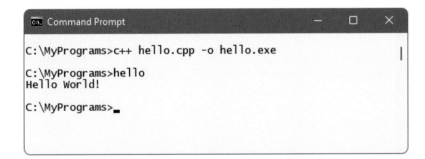

All command line examples in this book have been compiled and tested with the latest GNU C++ Compiler available at the time of writing – they may not replicate exactly with other compilers.

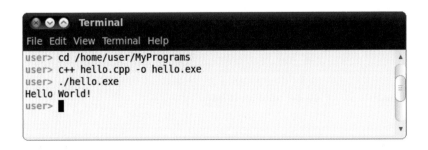

15

Creating variables

A "variable" is like a container in a C++ program in which a data value can be stored inside the computer's memory. The stored value can be referenced using the variable's name.

The programmer can choose any name for a variable, providing it adheres to the C++ naming conventions – a chosen name may only contain letters, digits, and the underscore character, but cannot begin with a digit. Also, the C++ keywords, listed on the inside cover of this book must be avoided. It's good practice to choose meaningful names to make the code more comprehensible.

To create a new variable in a program it must be "declared", specifying the type of data it may contain and its chosen name. A variable declaration has this syntax:

data-type variable-name ;

Multiple variables of the same data type can be created in a single declaration as a comma-separated list with this syntax:

data-type variable-name1 , variable-name2 , variable-name3 ;

The five basic C++ data types are listed in the table below, together with a brief description and example content:

Data Type	Description	Example
char	A single byte, capable of holding one character	'A'
int	An integer whole number	100
float	A floating-point number, correct to six decimal places	0.123456
double	A floating-point number, correct to 10 decimal places	0.0123456789
bool	A Boolean value of **true** or **false**, or numerically zero is false and any non-zero is true	**false** or 0 **true** or 1

Variable declarations must appear before executable statements – so they will be available for reference within statements.

Beware

Names are case-sensitive in C++ – so variables named **VAR**, **Var**, and **var** are treated as three individual variables. Traditionally, C++ variable names are lowercase and seldom begin with an underscore, as some C++ libraries use that convention.

Beware

Character values of the **char** data type must always be enclosed between single quotes – not double quotes.

When a value is assigned to a variable it is said to have been "initialized". Optionally, a variable may be initialized in its declaration. The value stored in any initialized variable can be displayed on standard output by the **cout** function, which was used on page 12 to display the "Hello World!" greeting.

vars.cpp

1. Start a new program by specifying the C++ library classes to include, and a namespace prefix to use
```
#include <iostream>
using namespace std ;
```

2. Add a main function containing a final **return** statement
```
int main()
{
  // Program code goes here.
  return 0 ;
}
```

3. In the main function, insert statements to declare and initialize variables of various data types
```
char letter ;     letter = 'A' ;      // Declared then initialized.
int number ;      number = 100 ;      // Declared then initialized.
float decimal = 7.5 ;                 // Declared and initialized.
double pi = 3.14159 ;                 // Declared and initialized.
bool isTrue = false ;                 // Declared and initialized.
```

4. Now, insert statements to output each stored value
```
cout << "char letter: "  << letter << endl ;
cout << "int number: " << number << endl ;
cout << "float decimal: " << decimal << endl ;
cout << "double pi: "   << pi << endl ;
cout << "bool isTrue: " << isTrue << endl ;
```

5. Save, compile, and run the program to see the output

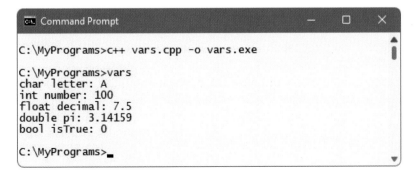

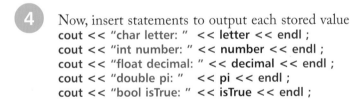

Hot tip

Always begin Boolean variable names with "is" so they are instantly recognizable as Booleans. Also, use "lowerCamelCase" for all variable names that comprise multiple words – where all except the first word begin with uppercase, like "isTrue".

17

Employing variable arrays

An array is a variable that can store multiple items of data – unlike a regular variable, which can only store one piece of data. The pieces of data are stored sequentially in array "elements" that are numbered, starting at 0. So, the first value is stored in element 0, the second value is stored in element 1, and so on.

An array is declared in the same way as other variables, but additionally the size of the array must also be specified in the declaration, in square brackets following the array name. For example, the syntax to declare an array named "nums" to store six integer numbers looks like this:

int nums[6] ;

Optionally, an array can be initialized when it is declared by assigning values to each element as a comma-separated list enclosed by curly brackets (braces). For example:

int nums[6] = { 0, 1, 2, 3, 4, 5 } ;

An individual element can be referenced using the array name followed by square brackets containing the element number. This means that **nums[1]** references the second element in the example above – not the first element, as element numbering starts at 0.

Arrays can be created for any C++ data type, but each element may only contain data of the same data type. An array of characters can be used to store a string of text if the final element contains the special **\0** null character. For example:

char name[5] = { 'm', 'i', 'k', 'e', '\0' } ;

The entire string to be referenced just by the array name. This is the principle means of working with strings in the C language, but the C++ string class, introduced in Chapter 4, is far simpler.

Collectively, the elements of an array are known as an "index". Arrays can have more than one index – to represent multiple dimensions, rather than the single dimension of a regular array. Multi-dimensional arrays of three indices and more are uncommon, but two-dimensional arrays are useful to store grid-based information, such as coordinates. For example:

int coords[2] [3] = { { 1, 2, 3 } , { 4, 5, 6 } } ;

Don't forget

Array numbering starts at 0 – so the final element in an array of six elements is number 5, not number 6.

	[0]	[1]	[2]
[0]	1	2	3
[1]	4	5	6

1. Start a new program by specifying the C++ library classes to include, and a namespace prefix to use

```cpp
#include <iostream>
using namespace std ;
```

arrays.cpp

2. Add a main function containing a final **return** statement

```cpp
int main()
{
  // Program code goes here.
  return 0 ;
}
```

Where possible, variable names should not be abbreviations – abbreviated names are only used in this book's examples due to space limitations.

3. In the main function, insert statements to declare and initialize three variable arrays

```cpp
// Declared then initialized.
float nums[3] ;
nums[0] = 1.5 ; nums[1] = 2.75 ; nums[2] = 3.25 ;

// Declared and initialized.
char name[5] = { 'm', 'i', 'k', 'e', '\0' } ;
int coords[2] [3] = { { 1, 2, 3 } , { 4, 5, 6 } } ;
}
```

4. Now, insert statements to output specific element values

```cpp
cout << "nums[0]: " << nums[0] << endl ;
cout << "nums[1]: " << nums[1] << endl ;
cout << "nums[2]: " << nums[2] << endl ;
cout << "name[0]: " << name[0] << endl ;
cout << "Text string: " << name << endl ;
cout << "coords[0][2]: " << coords[0][2] << endl ;
cout << "coords[1][2]: " << coords[1][2] << endl ;
```

The loop structures, introduced in Chapter 3, are often used to iterate array elements.

5. Save, compile, and run the program to see the output

```
C:\MyPrograms>c++ arrays.cpp -o arrays.exe

C:\MyPrograms>arrays
nums[0]: 1.5
nums[1]: 2.75
nums[2]: 3.25
name[0]: m
Text string: mike
coords[0][2]: 3
coords[1][2]: 6

C:\MyPrograms>_
```

19

Employing vector arrays

A vector is an alternative to a regular array, and has the advantage that its size can be changed as the program requires. Like regular arrays, vectors can be created for any data type, and their elements are also numbered starting at 0.

In order to use vectors in a program, the C++ **vector** library must be added with an **#include <vector>** preprocessor directive at the start of the program. This library contains the predefined functions in the table below, which are used to work with vectors:

Function:	Description:
at(*number*)	Gets the value contained in the specified element number
back()	Gets the value in the final element
clear()	Removes all vector elements
empty()	Returns true (1) if the vector is empty, or returns false (0) otherwise
front()	Gets the value in the first element
pop_back()	Removes the final element
push_back(*value*)	Adds a final element to the end of the vector, containing the specified value
size()	Gets the number of elements

A declaration to create a vector looks like this:

vector < *data-type* > *vector-name* (*size*) ;

An **int** vector will, by default, have each element automatically initialized with a zero value. Optionally, a different initial value can be specified after the size in the declaration, with this syntax:

vector < *data-type* > *vector-name* (*size* , *initial-value*) ;

The functions to work with vectors are simply appended to the chosen vector name by the dot operator. For example, to get the size of a vector named "vec" you would use **vec.size()**

Don't forget

Individual vector elements can be referenced using square brackets as with regular arrays, such as **vec[3]**

1 Start a new program by specifying the C++ library classes to include, and a namespace prefix to use

```cpp
#include <vector>              // Include vector support.
#include <iostream>
using namespace std ;
```

vector.cpp

2 Add a main function containing a final **return** statement

```cpp
int main()
{
  // Program code goes here.
  return 0 ;
}
```

3 In the main function, insert a statement to declare and initialize a vector array of three elements of the value 100

```cpp
vector <int> vec( 3, 100 ) ;
```

4 Now, insert statements to manipulate the vector elements

```cpp
cout << "Vector size: " << vec.size() << endl ;
cout << "Is empty?: " << vec.empty() << endl ;
cout << "First element: " << vec.at(0) << endl ;

vec.pop_back() ;          // Remove final element.
cout << "Vector size: " << vec.size() << endl ;
cout << "Final element: " << vec.back() << endl ;

vec.clear() ;             // Remove all elements.
cout << "Vector size: " << vec.size() << endl ;

vec.push_back( 200 ) ;  // Add an element.
cout << "Vector size: " << vec.size() << endl ;
cout << "First element: " << vec.front() << endl ;
```

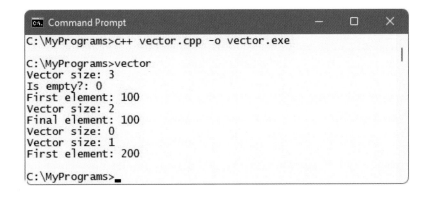

Hot tip

The example on page 50 shows how to use a loop to populate a vector with different initial values in each element.

5 Save, compile, and run the program to see the output

```
Command Prompt                           —    □    ✕
C:\MyPrograms>c++ vector.cpp -o vector.exe

C:\MyPrograms>vector
Vector size: 3
Is empty?: 0
First element: 100
Vector size: 2
Final element: 100
Vector size: 0
Vector size: 1
First element: 200

C:\MyPrograms>_
```

1000000

Don't
forget

The **typedef** keyword simply creates a nickname for a structure.

22

Declaring constants

Data that will not change during the execution of a program should be stored in a constant container, rather than in a variable. This better enables the compiler to check the code for errors – if the program attempts to change the value stored in a constant, the compiler will report an error and the compilation will fail.

A constant can be created for any data type by prefixing a variable declaration with the **const** keyword, followed by a space. Typically, constant names appear in uppercase to distinguish them from (lowercase) variable names. Unlike variables, constants must always be initialized in the declaration. For example, the declaration of a constant for the math pi value looks like this:

const double PI = 3.1415926536 ;

The **enum** keyword provides a handy way to create a sequence of integer constants in a concise manner. Optionally, the declaration can include a name for the sequence after the **enum** keyword. The constant names follow as a comma-separated list within braces. For example, this declaration creates a sequence of constants:

enum suit { CLUBS , DIAMONDS , HEARTS , SPADES } ;

Each of the constants will, by default, have a value one greater than the preceding constant in the list. Unless specified, the first constant will have a value of 0, the next a value of 1, and so on. A constant can be assigned any integer value, but the next constant in the list will always increment it by 1.

It is occasionally convenient to define a list of enumerated constants as a "custom data type" – by using the **typedef** keyword. This can begin the **enum** declaration, and a chosen type name can be added at the end of the declaration. For example, this **typedef** statement creates a custom data type named "charge":

typedef enum { NEGATIVE , POSITIVE } charge ;

Variables can then be created of the custom data type in the usual way, which may legally be assigned any of the listed constants. Essentially, these variables act just like an **int** variable – as they store the numerical integer value the assigned constant represents. For example, with the example above, assigning a **POSITIVE** constant to a **charge** variable actually assigns an integer of 1.

1 Start a new program by specifying the C++ library classes to include, and a namespace prefix to use
```
#include <iostream>
using namespace std ;
```

constant.cpp

2 Add a main function containing a final **return** statement
```
int main()
{
  // Program code goes here.
  return 0 ;
}
```

3 In the main function, insert statements to declare a constant, and output using the constant value
```
const double PI = 3.1415926536 ;
cout << "6\" circle circumference: " << (PI * 6) << endl ;
```

4 Next, insert statements to declare an enumerated list of constants, and output using some of those constant values
```
enum
{ RED=1, YELLOW, GREEN, BROWN, BLUE, PINK, BLACK } ;
cout << "I shot a red worth: " << RED << endl ;
cout << "Then a blue worth: " << BLUE << endl ;
cout << "Total scored: " << ( RED + BLUE ) << endl ;
```

Hot tip

In the PI declaration, the * character is the C++ multiplication operator, and the backslash character in \" escapes the quote mark from recognition – so the string does not get terminated prematurely.

5 Now, insert statements to declare a custom data type and output its assigned values
```
typedef enum { NEGATIVE , POSITIVE } charge ;
charge neutral = NEGATIVE , live = POSITIVE ;
cout << "Neutral wire: " << neutral << endl ;
cout << "Live wire: " << live << endl ;
```

6 Save, compile, and run the program to see the output

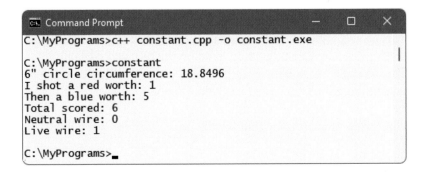

```
C:\MyPrograms>c++ constant.cpp -o constant.exe

C:\MyPrograms>constant
6" circle circumference: 18.8496
I shot a red worth: 1
Then a blue worth: 5
Total scored: 6
Neutral wire: 0
Live wire: 1

C:\MyPrograms>
```

Summary

- C++ is an object-oriented programming language that is an extension of the procedural C programming language.

- The GNU C++ Compiler is available for Windows and Linux.

- Preprocessor directives are used to make functions within the standard C++ libraries available to a program.

- Each C++ program must contain one main method as the entry point to the program.

- Statements define the actions that the program will execute.

- It is recommended that program code should be widely commented to make its purpose clear.

- The **c++** command calls the compiler, and its **-o** option allows the command to specify the name of the generated executable.

- A variable declaration specifies a data type and a chosen name by which the value within that variable can be referenced.

- The **cout** function, which is part of the C++ **iostream** library, writes content to the standard output console.

- An array is a fixed size variable that stores multiple items of data in elements, which are numbered starting at 0.

- The special **\0** character can be assigned to the final element of a **char** array to allow it to be treated as a single text string.

- A vector variable stores multiple items of data in elements, and can be dynamically resized.

- The value stored in an array or vector element can be referenced using that variable's name and its index number.

- Variable values that are never changed by the program should be stored in a constant.

- A constant list can be automatically numbered by the **enum** keyword and given a type name by the **typedef** keyword.

2 Performing operations

This chapter introduces

the C++ operators and

demonstrates the operations

they can perform.

Doing arithmetic

The arithmetical operators commonly used in C++ programs are listed in the table below, together with the operation they perform:

Operator:	Operation:
+	Addition
-	Subtraction
*	Multiplication
/	Division
%	Modulus
++	Increment
--	Decrement

The operators for addition, subtraction, multiplication, and division act as you would expect. Care must be taken, however, to bracket expressions where more than one operator is used to clarify the expression – operations within innermost parentheses are performed first:

a = b * c - d % e / f ;   // This is unclear.

a = (b * c) - ((d % e) / f) ;   // This is clearer.

The % modulus operator will divide the first given number by the second given number and return the remainder of the operation. This is useful to determine if a number has an odd or even value.

The ++ increment operator and -- decrement operator alter the given number by 1 and return the resulting value. These are most commonly used to count iterations in a loop. Counting up, the ++ operator increases the value by 1, while counting down, the -- decrement operator decreases the value by 1.

The increment and decrement operators can be placed before or after a value to different effect. If placed before the operand (prefix), its value is immediately changed; if placed after the operand (postfix), its value is noted first, then the value is changed.

Hot tip

Values used with operators to form expressions are called "operands" – in the expression 2 + 3 the numerical values 2 and 3 are the operands.

1 Start a new program by specifying the C++ library classes to include, and a namespace prefix to use

```
#include <iostream>
using namespace std ;
```

arithmetic.cpp

2 Add a main function containing a final **return** statement

```
int main()
{
  // Program code goes here.
  return 0 ;
}
```

3 In the main function, insert a statement to declare and initialize two integer variables

```
int a = 8 , b = 4 ;
```

4 Next, insert statements to output the result of each basic arithmetic operation

```
cout << "Addition result: "       << ( a + b ) << endl ;
cout << "Subtraction result: "    << ( a - b ) << endl ;
cout << "Multiplication result: " << ( a * b ) << endl ;
cout << "Division result: "       << ( a / b ) << endl ;
cout << "Modulus result: "        << ( a % b ) << endl ;
```

5 Now, insert statements to output the result of both postfix and prefix increment operations

```
cout << "Postfix increment: "  << a++ << endl ;
cout << "Postfix result: "     << a << endl ;
cout << "Prefix increment: "   << ++b << endl ;
cout << "Prefix result: "      << b << endl ;
```

Don't forget

A prefix operator changes the variable value immediately – a postfix operator changes the value subsequently.

6 Save, compile, and run the program to see the output

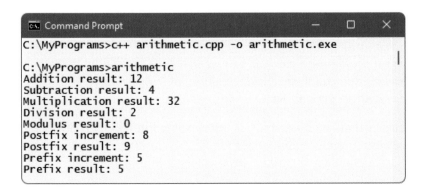

```
Command Prompt                              —    □    ×

C:\MyPrograms>c++ arithmetic.cpp -o arithmetic.exe

C:\MyPrograms>arithmetic
Addition result: 12
Subtraction result: 4
Multiplication result: 32
Division result: 2
Modulus result: 0
Postfix increment: 8
Postfix result: 9
Prefix increment: 5
Prefix result: 5
```

27

Assigning values

The operators that are used in C++ programming to assign values are listed in the table below. All except the simple = assignment operator are a shorthand form of a longer expression so each equivalent is given for clarity:

Operator:	Example:	Equivalent:
=	a = b	a = b
+=	a += b	a = (a + b)
-=	a -= b	a = (a - b)
*=	a *= b	a = (a * b)
/=	a /= b	a = (a / b)
%=	a %= b	a = (a % b)

In the example above, the variable named "a" is assigned the value that is contained in the variable named "b" – so *that* becomes the new value stored in the **a** variable.

The += operator is useful to add a value onto an existing value that is stored in the **a** variable.

In the table example, the += operator first adds the value contained in variable **a** to the value contained in variable **b**. It then assigns the result to become the new value stored in variable **a**.

All the other operators work in the same way by making the arithmetical operation between the two values first, then assigning the result of that operation to the first variable – to become its new stored value.

With the %= operator, the first operand **a** is divided by the second operand **b**, then the remainder of that operation is assigned to the **a** variable.

Each assignment operation is demonstrated in the program on the opposite page.

Don't forget

It is important to regard the = operator to mean "assign" rather than "equals" to avoid confusion with the == equality operator.

1 Start a new program by specifying the C++ library classes to include, and a namespace prefix to use
#include <iostream>
using namespace std ;

assign.cpp

2 Add a main function containing a final **return** statement
int main()
{
  // Program code goes here.
  return 0 ;
}

3 In the main function, insert a statement declaring two integer variables
int a , b ;

4 Next, insert statements to output simple assigned values
cout << "Assign values: " ;
cout << "a = " << (a = 8) << "   " ;
cout << "b = " << (b = 4) ;

5 Now, insert statements to output combined assigned values
cout << endl << "Add & assign: " ;
cout << "a += b (8 += 4) a = " << (a += b) ;
cout << endl << "Subtract & assign: " ;
cout << "a -= b (12 -= 4) a = " << (a -= b) ;
cout << endl << "Multiply & assign: " ;
cout << "a *= b (8 *= 4) a = " << (a *= b) ;
cout << endl << "Divide & assign: " ;
cout << "a /= b (32 /= 4) a = " << (a /= b) ;
cout << endl << "Modulus & assign: " ;
cout << "a %= b (8 %= 4) a = " << (a %= b) ;

Unlike the = assign operator, the == equality operator compares operands and is described on page 30.

6 Save, compile, and run the program to see the output

```
Command Prompt                           —    □    ×

C:\MyPrograms>c++ assign.cpp -o assign.exe

C:\MyPrograms>assign
Assign values:  a = 8    b = 4
Add & assign:  a += b (8 += 4 ) a = 12
Subtract & assign: a -= b (12 -= 4 ) a = 8
Multiply & assign: a *= b (8 *= 4 ) a = 32
Divide & assign: a /= b (32 /= 4 ) a = 8
Modulus & assign: a %= b (8 %= 4 )  a = 0
```

Comparing values

The operators that are commonly used in C++ programming to compare two numerical values are listed in the table below:

Operator:	Comparative test:
==	Equality
!=	Inequality
>	Greater than
<	Less than
>=	Greater than or equal to
<=	Less than or equal to

Hot tip

A-Z uppercase characters have ASCII code values 65-90, and a-z lowercase characters have ASCII code values 97-122.

The == equality operator compares two operands and will return **true** (1) if both are equal in value, otherwise it will return a **false** (0) value. If both are the same number, they are equal, or if both are characters, their ASCII code values are compared numerically. Conversely, the != inequality operator returns **true** (1) if two operands are not equal, using the same rules as the == equality operator, otherwise it returns **false** (0). Equality and inequality operators are useful in testing the state of two variables to perform conditional branching in a program.

The > "greater than" operator compares two operands and will return **true** (1) if the first is greater in value than the second, or it will return **false** (0) if it is equal or less in value. The < "less than" operator makes the same comparison but returns **true** (1) if the first operand is less in value than the second, otherwise it returns **false** (0). A > "greater than" or < "less than" operator is often used to test the value of an iteration counter in a loop.

Adding the = operator after a > "greater than" or < "less than" operator makes it also return **true** (1) if the two operands are exactly equal in value.

Each comparison operation is demonstrated in the program on the opposite page.

...cont'd

1 Start a new program by specifying the C++ library classes to include and a namespace prefix to use

```
#include <iostream>
using namespace std ;
```

comparison.cpp

2 Add a main function containing a final **return** statement

```
int main()
{
  // Program code goes here.
  return 0 ;
}
```

3 In the main function, insert statements to declare and initialize variables that can convert to Booleans

```
int nil = 0, num = 0, max = 1 ; char cap = 'A', low = 'a' ;
```

4 Next, insert statements to output equality comparisons of integers and characters

```
cout << "Equality comparisons: " ;
cout << "(0 == 0) " << ( nil == num ) << "(true)" ;
cout << "(A == a) " << ( cap == low ) << "(false)" ;
```

5 Now, insert statements to output all other comparisons

```
cout << endl << "Inequality comparison: " ;
cout << "(0 != 1) " << ( nil != max )   << "(true)" ;
cout << endl << "Greater comparison: " ;
cout << "(0 > 1) " << ( nil > max )      << "(false)" ;
cout << endl << "Lesser comparison: " ;
cout << "(0 < 1) " << ( nil < max )      << "(true)" ;
cout << endl << "Greater or equal comparison: " ;
cout << "(0 >= 0) "<< ( nil >= num ) << "(true)" ;
cout << endl << "Lesser or equal comparison: " ;
cout << "(1 <= 0) " << ( max <= num ) << "(false)" ;
```

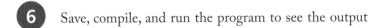

Don't forget

The ASCII code value for uppercase "A" is 65, but for lowercase "a" it's 97 – so their comparison here returns **false** (0).

6 Save, compile, and run the program to see the output

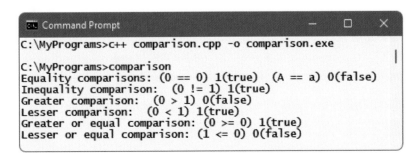

```
Command Prompt                              —    □    ✕

C:\MyPrograms>c++ comparison.cpp -o comparison.exe

C:\MyPrograms>comparison
Equality comparisons: (0 == 0) 1(true)  (A == a) 0(false)
Inequality comparison:  (0 != 1) 1(true)
Greater comparison:  (0 > 1) 0(false)
Lesser comparison:  (0 < 1) 1(true)
Greater or equal comparison: (0 >= 0) 1(true)
Lesser or equal comparison: (1 <= 0) 0(false)
```

Assessing logic

The logical operators most commonly used in C++ programming are listed in the table below:

Operator:	Operation:
&&	Logical AND
||	Logical OR
!	Logical NOT

The logical operators are used with operands that have Boolean values of **true** or **false**, or are values that convert to **true** or **false**.

The logical **&&** AND operator will evaluate two operands and return **true** only if both operands themselves are **true**. Otherwise, the **&&** operator will return **false**. This is used in conditional branching where the direction of a program is determined by testing two conditions – if both conditions are satisfied, the program will go in a certain direction, otherwise it will take a different direction.

Unlike the **&&** AND operator that needs both operands to be **true**, the **||** OR operator will evaluate its two operands and return **true** if either one of the operands itself returns **true**. If neither operand returns **true** then the **||** OR operator will return **false**. This is useful in C++ programming to perform a certain action if either one of two test conditions has been met.

The third logical **!** NOT operator is a unary operator that is used before a single operand. It returns the inverse value of the given operand, so if the variable **a** had a value of **true**, then **!a** would have a value of **false**. The **!** NOT operator is useful in C++ programs to toggle the value of a variable in successive loop iterations with a statement like **a = !a**. This ensures that on each pass the value is changed, like flicking a light switch on and off.

In C++ programs, a 0 represents the Boolean **false** value and any non-zero value, such as 1, represents the Boolean **true** value.

Each logical operation is demonstrated in the program on the opposite page.

Beware

Where there is more than one operand, each expression must be enclosed by parentheses.

The term "Boolean" refers to a system of logical thought developed by the English mathematician George Boole (1815-1864).

32

1 Start a new program by specifying the C++ library classes to include, and a namespace prefix to use
#include <iostream>
using namespace std ;

logic.cpp

2 Add a main function containing a final **return** statement
int main()
{
  // Program code goes here.
  return 0 ;
}

3 In the main function, declare and initialize two integer variables – with values that can represent Boolean values
int a = 1 , b = 0 ;

4 Insert statements to output the result of AND evaluations
cout << "AND logic:" << endl ;
cout << "(a && a) " << (a && a) << "(true) " ;
cout << "(a && b) " << (a && b) << "(false) " ;
cout << "(b && b) " << (b && b) << "(false)" << endl ;

5 Insert statements to output the result of OR evaluations
cout << endl << "OR logic:" << endl ;
cout << "(a || a) " << (a || a) << "(true) " ;
cout << "(a || b) " << (a || b) << "(true) " ;
cout << "(b || b) " << (b || b) << "(false)" << endl ;

6 Insert statements to output the result of NOT evaluations
cout << endl << "NOT logic:" << endl ;
cout << "a = " << a << " !a = " << !a << "   " ;
cout << "b = " << b << " !b = " << !b << endl ;

7 Save, compile and run the program to see the output

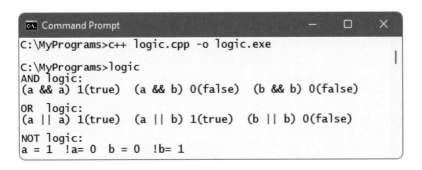

33

Don't forget

Notice that **0 && 0** returns **0**, not **1** – demonstrating the maxim "two wrongs don't make a right".

Examining conditions

Possibly the C++ programmer's most favorite test operator is the ?: "ternary" operator. This operator first evaluates an expression for a **true** or **false** condition, then returns one of two specified values depending on the result of the evaluation. For this reason it is also known as the "conditional" operator.

The **?:** ternary operator has this syntax:

(*test-expression*) ? *if-true-return-this* : *if-false-return-this* ;

Although the ternary operator can initially appear a little confusing, it is well worth becoming familiar with this operator as it can execute powerful program branching with minimal code. For example, to branch when a variable is not a value of 1:

(var != 1) ? *if-true-do-this* : *if-false-do-this* ;

The ternary operator is commonly used in C++ programming to assign the maximum or minimum value of two variables to a third variable. For example, to assign a minimum like this:

c = (a < b) ? a : b ;

The expression in parentheses returns **true** when the value of variable **a** is less than that of variable **b** – so in this case, the lesser value of variable **a** gets assigned to variable **c**.

Similarly, replacing the **<** less than operator in the test expression with the **>** greater than operator would assign the greater value of variable **b** to variable **c**.

Another common use of the ternary operator incorporates the **%** modulus operator in the test expression to determine whether the value of a variable is an odd number or an even number:

(var % 2 != 0) ? *if-true(odd)-do-this* : *if-false(even)-do-this* ;

Where the result of dividing the variable value by two does leave a remainder, the number is odd – where there is no remainder, the number is even. The test expression (**var % 2 == 1**) would have the same effect but it is preferable to test for inequality – it's easier to spot when something is different than when it's identical.

The ternary operator is demonstrated in the program on the opposite page.

Don't forget

The ternary operator has three operands – the one before the **?**, and those before and after the **:**.

1 Start a new program by specifying the C++ library classes to include, and a namespace prefix to use

```cpp
#include <iostream>
using namespace std ;
```

ternary.cpp

2 Add a main function containing a final **return** statement

```cpp
int main()
{
  // Program code goes here.
  return 0 ;
}
```

3 In the main function, insert statements declaring three integer variables, and initializing two of them

```cpp
int a, b, max ;
a = 1, b = 2 ;
```

4 Insert statements to output the value and parity of the first examined variable

```cpp
cout << "Variable a value is: " ;
cout << ( ( a != 1 ) ? "not 1, " : "1, " ) ;
cout << ( ( a % 2 != 0 ) ? "odd" : "even" ) ;
```

5 Next, insert statements to output the value and parity of the second examined variable

```cpp
cout << endl << "Variable b value is: " ;
cout << ( ( b != 1 ) ? "not 1, " : "1, " ) ;
cout << ( ( b % 2 != 0 ) ? "odd" : "even" ) ;
```

6 Now, insert statements to output the greater of the two stored variable values

```cpp
max = ( a > b ) ? a : b ;
cout << endl << "Greater value is: " << max << endl ;
```

7 Save, compile and run the program to see the output

Hot tip

The ternary operator can return values of any data type – numbers, strings, Boolean values, etc.

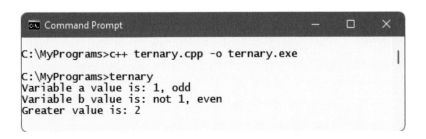

```
Command Prompt                        —    □    ×

C:\MyPrograms>c++ ternary.cpp -o ternary.exe     |

C:\MyPrograms>ternary
Variable a value is: 1, odd
Variable b value is: not 1, even
Greater value is: 2
```

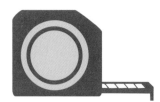

Establishing size

Declaration of a variable allocates system memory where values assigned to that variable will be stored. The amount of memory allocated for this is determined by your system and the data type.

Typically, an **int** data type is created as a "long" value by default, which can store values from +2,147,483,647 to -2,147,483,648. On the other hand, if the **int** data type is created as a "short" value by default, it can only store values from +32,767 to -32,768.

The preferred range can be explicitly specified when declaring the variable by prefixing the **int** keyword with a **short** or **long** qualifier. The **short int** is useful to save memory space when you are sure the limited range will never be exceeded.

When an **int** variable is declared, it can by default contain either positive or negative integers, which are known as "signed" values. If the variable will always contain only positive integers, it can be qualified as **unsigned** to increase its maximum possible value. Typically, an **unsigned short int** has a range from zero to 65,535 and an **unsigned long int** has a range from zero to 4,294,967,295.

The memory size of any variable can be discovered using the C++ **sizeof** operator. The name of the variable to be examined can be specified in optional parentheses following the **sizeof** operator name. For example, to examine a variable named "var":

sizeof(var) ;   // Alternatively you can use "sizeof var ;".

The **sizeof** operator will return an integer that is the number of bytes allocated to store data within the named variable.

Simple data types, such as **char** and **bool**, only need a single byte of memory to store just one piece of data. Longer numeric values need more memory, according to their possible range – determined by data type and qualifiers.

The memory allocated to an array is simply a multiple of that allocated to a single variable of its data type, according to its number of elements. For example, an **int** array of 50 elements will allocate 50 times the memory allocated to a single **int** variable.

The **sizeof** operator is demonstrated in the program on the opposite page.

Hot tip

Although **sizeof** is an operator that does not strictly need parentheses, it is commonly seen with them – as if it was a function, like **main()**

...cont'd

1 Start a new program by specifying the C++ library classes to include, and a namespace prefix to use

```
#include <iostream>
using namespace std ;
```

sizeof.cpp

2 Add a main function containing a final **return** statement

```
int main()
{
  // Program code goes here.
  return 0 ;
}
```

3 In the main function, insert statements declaring variables of various data types

```
int num ;          int nums[50] ;         float decimal ;
bool isTrue ;      unsigned int max ;     char letter ;
double pi ;        short int number ;     char letters[50] ;
```

4 Next, insert statements to output the byte size of each integer variable

```
cout << "int size:" << sizeof( num ) << endl ;
cout << "50 int size: " << sizeof( nums ) << endl ;
cout << "short int size: " << sizeof( number ) << endl ;
cout << "unsigned int size: " << sizeof( max ) << endl ;
```

5 Now, insert statements to output the size of other variables

```
cout << "double size: " << sizeof( pi ) << endl ;
cout << "float size: " << sizeof( decimal ) << endl ;
cout << "char size: " << sizeof( letter ) << endl ;
cout << "50 char size: " << sizeof( letters ) << endl ;
cout << "bool size: " << sizeof( isTrue ) << endl ;
```

Here, the **int** data type is created as a **long** type by default – your system may be different.

6 Save, compile and run the program to see the output

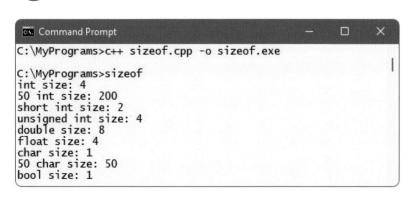

```
C:\MyPrograms>c++ sizeof.cpp -o sizeof.exe

C:\MyPrograms>sizeof
int size: 4
50 int size: 200
short int size: 2
unsigned int size: 4
double size: 8
float size: 4
char size: 1
50 char size: 50
bool size: 1
```

37

Setting precedence

Operator precedence determines the order in which C++ evaluates expressions. For example, in the expression **a = 6 + 8 * 3**, the order of precedence determines that multiplication is completed first.

The table below lists operator precedence in descending order – those on the top row have highest precedence; those on lower rows have successively lower precedence. The order in which C++ evaluates expressions containing multiple operators of equal precedence is determined by "operator associativity" – grouping operands with the one on the left (LTR) or on the right (RTL).

Don't forget

The * multiply operator is on a higher row than the + addition operator – so in the expression **a=6+8*3**, multiplication is completed first, before the addition.

Hot tip

The **->** class pointer and the **.** class member operators are introduced later in this book – but they are included here for completeness.

Operator:				Direction:
( )	Function call	[ ]	Array index	LTR
->	Class pointer	.	Class member	
!	Logical NOT	*	Pointer	RTL
--	Decrement	++	Increment	
+	Positive sign	-	Negative sign	
sizeof	Size of	&	Address of	
*	Multiply	/	Divide	LTR
%	Modulus			
+	Add	-	Subtract	LTR
<=	Less or equal	<	Less than	LTR
>=	Greater or equal	>	Greater than	
==	Equality	!=	Inequality	LTR
&&	Logical AND			LTR
||	Logical OR			LTR
?:	Ternary			RTL
+= -= *= /= %= Assignments				RTL
,	Comma			LTR

In addition to the operators in this table there are a number of "bitwise" operators, which are used to perform binary arithmetic. This is outside the scope of this book, but there is a section devoted to binary arithmetic in **C Programming in easy steps**. Those operators perform in just the same way in C++.

Operator precedence is demonstrated in the program opposite.

1 Start a new program by specifying the C++ library classes to include, and a namespace prefix to use

```
#include <iostream>
using namespace std ;
```

precedence.cpp

2 Add a main function continuing a final **return** statement

```
int main()
{
   // Program code goes here.
   return 0 ;
}
```

3 In the main function, declare an integer variable initialized with the result of an expression using default precedence, then output the result

```
int num = 1 + 4 * 3 ;
cout << endl << "Default order: " << num << endl ;
```

4 Next, assign the result of this expression to the variable using explicit precedence, then output the result

```
num = ( 1 + 4 ) * 3 ;
cout << "Forced order: " << num << endl << endl ;
```

5 Assign the result of a different expression to the variable using direction precedence, then output the result

```
num = 7 - 4 + 2 ;
cout<< "Default direction: " << num << endl ;
```

6 Now, assign the result of this expression to the variable using explicit precedence, then output the result

```
num = 7 - ( 4 + 2 ) ;
cout << "Forced direction: " << num << endl ;
```

Beware

Do not rely upon default precedence as it may vary between compilers – always use parentheses to clarify expressions.

7 Save, compile and run the program to see the output

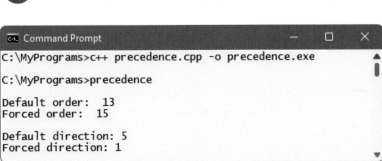

```
C:\MyPrograms>c++ precedence.cpp -o precedence.exe

C:\MyPrograms>precedence

Default order:   13
Forced order:   15

Default direction: 5
Forced direction: 1
```

39

Casting data types

Any data stored in a variable can be forced (coerced) into a variable of a different data type by a process known as "casting". The cast statement simply states the data type to which the value should be cast in parentheses preceding the name of the variable containing the data to be cast. So casting syntax looks like this:

variable-name = (*data-type*) *variable-name* ;

This is the traditional form of casting that is also found in the C programming language. A newer alternative available in C++ uses angled brackets with the **static_cast** keyword, like this:

variable-name = static_cast < *data-type* > *variable-name* ;

The newer version allows casts to be more easily identified in source code by avoiding the use of parentheses, which can easily be confused with parentheses in expressions. The newer form of casting is preferred, but the older form is still widely found.

Casting is often necessary to accurately store the result of an arithmetic operation, because dividing one integer by another integer will always produce an integer result. For example, the integer division **7/2** produces the truncated integer result of **3**.

To store the accurate floating-point result would require the result be cast into a suitable data type, such as a **float**, like this:

float result = (float) 7 / 2 ;

Or alternatively using the newer form of cast:

float result = static_cast < float > 7 / 2 ;

In either case, it should be noted that operator precedence casts the first operand into the specified data type before implementing the arithmetic operation, so the statement can best be written as:

float result = static_cast < float > (7) / 2 ;

Bracketing the expression as **(7 / 2)** would perform the arithmetic first on integers, so the integer result would be truncated before being cast into the **float** variable – not the desired effect!

Casting with both the older C-style form and the newer C++ form is demonstrated in the program on the opposite page.

Don't forget

The result of dividing an integer by another integer is truncated, not rounded – so a result of 9.9 would become 9.

40

…cont'd

1 Start a new program by specifying the C++ library classes to include, and a namespace prefix to use

```
#include <iostream>
using namespace std ;
```

cast.cpp

2 Add a main function containing a final **return** statement

```
int main()
{
  // Program code goes here.
  return 0 ;
}
```

3 In the main function, insert statements to declare and initialize integer, character, and floating-point variables

```
int num = 7, factor = 2 ;
char letter = 'A' ; float result = 0.0 ;
```

4 Output the result of a plain integer division

```
cout << "Integer division: " << ( num / factor ) << endl ;
```

5 Now, cast the same division into a floating-point variable and output that result

```
result = (float) ( num ) / factor ;
cout << "Cast division float: " << result << endl ;
```

6 Next, cast a character variable into an integer variable and output that value

```
num = static_cast <int> ( letter ) ;
cout << "Cast character int: " << num << endl ;
```

7 Cast an integer into a character variable and output it

```
letter = static_cast <char> ( 70 ) ;
cout << "Cast integer char: " << letter << endl ;
```

8 Save, compile and run the program to see the output

Hot tip

ASCII (pronounced "askee") is the American Standard Code for Information Interchange, which is the accepted standard for plain text. In ASCII, characters are represented numerically within the range 0-127. Uppercase 'A' is 65, so that integer value gets cast into an **int** variable.

41

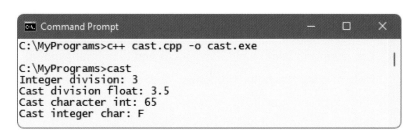

```
C:\MyPrograms>c++ cast.cpp -o cast.exe

C:\MyPrograms>cast
Integer division: 3
Cast division float: 3.5
Cast character int: 65
Cast integer char: F
```

Summary

- Arithmetical operators can form expressions with two operands for addition +, subtraction -, multiplication *, division /, or modulus %.

- Increment ++ and decrement -- operators modify a single operand by a value of 1.

- The assignment = operator can be combined with an arithmetical operator to perform an arithmetical calculation, then assign its result.

- Comparison operators can form expressions comparing two operands for equality ==, inequality !=, greater >, lesser <, greater or equal >=, and lesser or equal <= values.

- Logical && and || operators form expressions evaluating two operands to return a Boolean value of true or false.

- The logical ! operator returns the inverse Boolean value of a single operand.

- A ternary ?: operator evaluates a given Boolean expression, then returns one of two operands depending on its result.

- The **sizeof** operator returns the memory byte size of a variable.

- An **int** variable may be qualified as a **short** type for smaller numbers, or as a **long** type for large numbers.

- Where an **int** variable will only store positive numbers, it may be qualified as **unsigned** to extend its numeric range.

- It is important to explicitly set operator precedence in complex expressions by adding parentheses ( ).

- Data stored in a variable can be forced into a variable of a different data type by the casting process.

- C++ supports traditional C-style casts and the newer form of casts that use the **static_cast** keyword.

3 Making statements

This chapter demonstrates

C++ conditional statements,

which allow programs

to branch in different

directions, and introduces

C++ function structures.

Branching with if

The C++ **if** keyword performs the basic conditional test that evaluates a given expression for a Boolean value of **true** or **false** – and its syntax looks like this:

if (*test-expression*) { *statements-to-execute-when-true* }

The braces following the test may contain one or more statements, each terminated by a semi-colon, but these will only be executed when the expression is found to be **true**. When the test is found to be **false**, the program proceeds to its next task.

Optionally, an **if** statement can offer alternative statements to execute when the test fails by appending an **else** statement block after the **if** statement block, like this:

if (*test-expression*) { *statements-to-execute-when-true* }
else { *statements-to-execute-when-false* }

To test two conditions, the test expression may use the **&&** operator – for example, **if ((num > 5) && (letter == 'A'))**. Alternatively, an **if** statement can be "nested" within another **if** statement, so those statements in the inner statement block will only be executed when both tests succeed – but statements in the outer statement block will be executed if the outer test succeeds.

Hot tip

Where there is only one statement to execute when the test succeeds, the braces may be omitted – but retaining them aids code clarity.

ifelse.cpp

1 Start a new program by specifying the C++ library classes to include, and a namespace prefix to use
```
#include <iostream>
using namespace std ;
```

2 Add a main function containing a final **return** statement
```
int main()
{
  // Program code goes here.
  return 0 ;
}
```

3 In the main function, insert statements to declare and initialize two variables
```
int num = 8 ;
char letter = 'A' ;
```

④ Next, insert an **if-else** statement that tests the integer variable value and outputs an appropriate response
```
if ( num > 5 )
{ cout << "Number exceeds five" << endl ; }
else
{ cout << "Number is five or less" << endl ; }
```

⑤ In the **if** statement block, insert a nested **if** statement that tests the character variable value and outputs when matched
```
if ( letter == 'A' ) { cout << "Letter is A" << endl ; }
```

Hot tip

Shorthand can be used when testing a Boolean value – so the expression **if (flag == true)** can be written as **if (flag)**

⑥ Save, compile and run the program to see both tests succeed

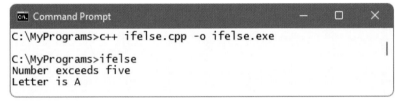

```
C:\MyPrograms>c++ ifelse.cpp -o ifelse.exe

C:\MyPrograms>ifelse
Number exceeds five
Letter is A
```

⑦ Edit the character variable declaration to change its value
```
char letter = 'B' ;
```

⑧ Save, compile, and run the program once more to see only the outer test succeed – executing the outer **if** statement

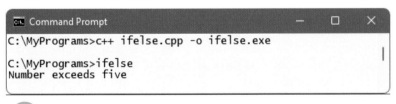

```
C:\MyPrograms>c++ ifelse.cpp -o ifelse.exe

C:\MyPrograms>ifelse
Number exceeds five
```

Beware

Avoid nesting more than three levels of **if** statements – to avoid confusion and errors.

⑨ Edit the integer variable declaration to change its value
```
int num = 3 ;
```

⑩ Save, compile, and run the program again to see both tests now fail – executing the outer **else** statement

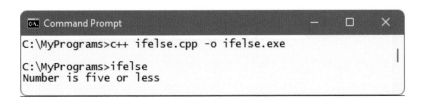

```
C:\MyPrograms>c++ ifelse.cpp -o ifelse.exe

C:\MyPrograms>ifelse
Number is five or less
```

Switching branches

The **if** and **else** keywords, introduced on pages 44-45, allow programs to branch in a particular direction according to the result of a test condition, and can be used to repeatedly test a variable to match a value – for example, testing for an integer:

```
if ( num == 1 ) { cout << "Monday" ; }
else
if ( num == 2 ) { cout << "Tuesday" ; }
else
if ( num == 3 ) { cout << "Wednesday" ; }
else
if ( num == 4 ) { cout << "Thursday" ; }
else
if ( num == 5 ) { cout << "Friday" ; }
```

The program will branch in the direction of the match.

Conditional branching with long **if-else** statements can often be more efficiently performed using a **switch** statement instead, especially when the test expression evaluates one variable.

The **switch** statement works in an unusual way. It takes a given variable value, then seeks a matching value among a number of **case** statements. Statements associated with the matching case statement value will then be executed.

When no match is found, no **case** statements will be executed, but you may add a **default** statement after the final **case** statement to specify statements to be executed when no match is found.

Beware

It is important to follow each case statement with the **break** keyword, to stop the program proceeding through the switch block after all statements associated with the matched **case** value have been executed – unless that is precisely what you require. For example, one statement for each block of three values, like this:

```
switch( variable-name )
{
  case value1 ; case value2 ; case value3 ;
       statements-to-be-executed ; break ;

case value4 ; case value5 ; case value6 ;
       statements-to-be-executed ; break ;
}
```

Usually, each **case** statement will have its own set of statements to execute and be terminated by a **break**, as in the program opposite.

1 Start a new program by specifying the C++ library classes to include, and a namespace prefix to use

```
#include <iostream>
using namespace std ;
```

switch.cpp

2 Add a main function containing a final **return** statement

```
int main()
{
  // Program code goes here.
  return 0 ;
}
```

3 In the main function, insert a statement to declare and initialize an integer variable with a value to be matched

```
int num = 3 ;
```

4 Next, insert a **switch** statement to seek a match

```
switch ( num )
{
  case 1 : cout << num << " : Monday" ; break ;
  case 2 : cout << num << " : Tuesday" ; break ;
  case 3 : cout << num << " : Wednesday" ; break ;
  case 4 : cout << num << " : Thursday" ; break ;
  case 5 : cout << num << " : Friday" ; break ;
}
```

5 In the **switch** statement, insert a **default** statement after the final **case** statement

```
default : cout << num << " : Weekend day" ;
```

6 Save, compile, and run the program to see the output

7 Now, edit the integer variable declaration to change its value, then save, compile and run the program once more

```
int num = 6 ;
```

> **Hot tip**
>
> Notice that a **default** statement does not need to be followed by a **break** keyword – because a **default** statement always appears last in a **switch** statement.

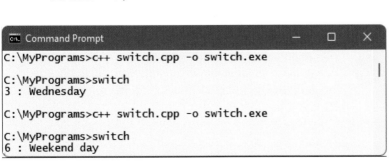

```
C:\MyPrograms>c++ switch.cpp -o switch.exe

C:\MyPrograms>switch
3 : Wednesday

C:\MyPrograms>c++ switch.cpp -o switch.exe

C:\MyPrograms>switch
6 : Weekend day
```

Looping for

A loop is a piece of code in a program that automatically repeats. One complete execution of all statements contained within the loop block is known as an "iteration" or "pass".

The number of iterations made by a loop is controlled by a conditional test made within the loop. While the tested expression remains **true**, the loop will continue – until the tested expression becomes **false**, at which time the loop ends.

The three types of loop structures in C++ programming are **for** loops, **while** loops, and **do-while** loops. Perhaps the most commonly used loop is the **for** loop, which has this syntax:

for (*initializer* ; *test-expression* ; *incrementer*) { *statements* }

The initializer sets the starting value for a counter of the number of iterations made by the loop. An integer variable is used for this purpose and is traditionally named "i".

Upon each iteration of the loop, the test expression is evaluated, and that iteration will only continue while this expression is **true**. When the tested expression becomes **false**, the loop ends immediately without executing the statements again. On each iteration the counter is incremented then the statements executed.

Loops may be nested within other loops – so that the inner loop will fully execute its iterations on each iteration of the outer loop.

forloop.cpp

1 Start a new program by specifying the C++ library classes to include, and a namespace prefix to use
   #include <iostream>
   using namespace std ;

2 Add a main function containing a final **return** statement
   int main()
   {
     // Program code goes here.
     return 0 ;
   }

3 In the main function, insert a statement to declare an integer variable to be used as a loop iteration counter
   int i ;

...cont'd

4 Next, insert a **for** loop to output the counter value on each of three iterations
```
for ( i = 1 ; i < 4 ; i++ )
{
  cout << "Loop iteration: " << i << endl ;
}
```

Hot tip

Alternatively, a **for** loop counter can count down by decrementing the counter value on each iteration using i-- instead of the i++ incrementer.

5 Save, compile, and run the program to see the output

```
C:\MyPrograms>c++ forloop.cpp -o forloop.exe

C:\MyPrograms>forloop
Loop iteration: 1
Loop iteration: 2
Loop iteration: 3

C:\MyPrograms>
```

6 Now, edit the variable declaration to add a second counter
```
int i , j ;                    // Integer variable "j" added.
```

7 Inside the **for** loop block, after the output statement add an inner loop to output its counter value on each iteration
```
for ( j = 1 ; j < 4 ; j++ )
{ cout << "    Inner loop iteration: " << j << endl ; }
```

8 Save, compile, and run the program again to see the inner loop fully execute on each iteration of the outer loop

```
C:\MyPrograms>c++ forloop.cpp -o forloop.exe

C:\MyPrograms>forloop
Loop iteration: 1
        Inner loop iteration: 1
        Inner loop iteration: 2
        Inner loop iteration: 3
Loop iteration: 2
        Inner loop iteration: 1
        Inner loop iteration: 2
        Inner loop iteration: 3
Loop iteration: 3
        Inner loop iteration: 1
        Inner loop iteration: 2
        Inner loop iteration: 3
```

Don't forget

On the third iteration of these loops, the incrementer increases the counter value to 4 – so when it is next evaluated, the test returns false and the loop ends.

Looping while

An alternative to the **for** loop, introduced on pages 48-49, uses the **while** keyword, followed by an expression to be evaluated. When the expression is **true**, statements contained within braces following the test expression will be executed. The expression will then be evaluated again, and the **while** loop will continue until the expression is found to be **false**.

The loop's statement block <u>must</u> contain code that will affect the tested expression in order to change the evaluation result to **false**, otherwise an infinite loop is created that will lock the system! When the tested expression is found to be **false** upon its first evaluation, the **while** loop's statement block will never be executed.

A subtle variation of the **while** loop places the **do** keyword before the loop's statement block and a **while** test after it, with this syntax:

do { statements-to-be-executed } while (test-expression) ;

In a **do-while** loop, the statement block will always be executed at least once – because the expression is not evaluated until after the first iteration of the loop.

A **break** statement can be included in any kind of loop to immediately terminate the loop when a test condition is met. The **break** statement ensures no further iterations of that loop will be executed.

Similarly, a **continue** statement can be included in any kind of loop to immediately terminate that particular iteration of the loop when a test condition is met. The **continue** statement allows the loop to proceed to the next iteration.

1️⃣ Start a new program by specifying the C++ library classes to include, and a namespace prefix to use

```
#include <vector>          // Include vector support.
#include <iostream>
using namespace std ;
```

2️⃣ Add a main function containing a final **return** statement

```
int main()
{
  // Program code goes here.
  return 0 ;
}
```

C++

while.cpp

...cont'd

3 In the main function, insert statements to declare an integer vector and an integer variable loop counter

```
vector <int> vec( 10 ) ;
int i = 0 ;
```

The vector library must be included with a preprocessor directive in this example.

4 Next, insert a **while** loop to assign a counter value to an element of the vector on each iteration

```
while ( i < vec.size() )
{
  i++ ;                    // Increment the counter.
  vec[ i-1 ] = i ;         // Assign count to element.
  cout << " | " << vec.at( i-1 ) ;
}
```

5 Save compile and run the program to see the output

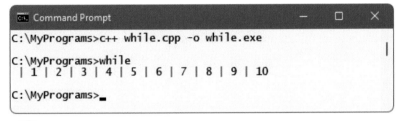

```
C:\MyPrograms>c++ while.cpp -o while.exe

C:\MyPrograms>while
 | 1 | 2 | 3 | 4 | 5 | 6 | 7 | 8 | 9 | 10

C:\MyPrograms>
```

51

6 Edit the **while** loop to add a **continue** statement immediately after the incrementer to make the loop skip its third iteration

```
if ( i == 3 ) { cout << " | Skipped" ; continue ; }
```

The position of **break** and **continue** statements is important – they must appear after the incrementer, to avoid creating an infinite loop, but before other statements that affect the program to avoid executing those statements.

7 After the **continue** statement, now add a **break** statement, to make the loop quit on its eighth iteration

```
if ( i == 8 ) { cout << endl << "Done" ; break ; }
```

8 Save, compile, and run the program once more to see the loop now omits some iterations

```
C:\MyPrograms>c++ while.cpp -o while.exe

C:\MyPrograms>while
 | 1 | 2 | Skipped | 4 | 5 | 6 | 7
Done

C:\MyPrograms>
```

Declaring functions

Functions enclose a section of code that provides specific functionality to the program. When a function is called from the main program, its statements are executed and, optionally, a value can be returned to the main program upon completion. There are three main benefits to using functions:

- Functions make program code easier to understand and maintain.

- Tried and tested functions can be re-used by other programs.

- Several programmers can divide the workload in large projects by working on different functions of the program.

Declaring functions

Each function is declared early in the program code as a "prototype", comprising a data type for the value it will return and the function name followed by parentheses, which may optionally contain a list of "argument" data types of passed values it may use. The syntax of a function prototype declaration looks like this:

return-data-type function-name (*arguments-data-type-list*) ;

For example, a function named "computeArea" that returns a **float** value and is passed two **float** arguments is declared as:

float computeArea(float, float) ;

Defining functions

The function's definition appears later in the program code and comprises a repeat of the prototype, plus the actual function body. The function body is the statements to be executed whenever the function is called, contained within a pair of braces.

It is important to recognize that the compiler checks the function definition against the prototype, so the actual returned data type must match that specified in the prototype, and any supplied arguments must match in both number and data type. Compilation fails if the definition does not match the prototype. A simple **computeArea** definition might look like this:

```
float computeArea( float width, float height )
{
        return ( width * height ) ;
}
```

Strictly speaking, the arguments in a function prototype are known as its "formal parameters".

Use the **void** keyword if the function will return no value to the caller.

Variable scope

Variables that are declared in a function can only be used locally within that function, and are not accessible globally for use in other functions. This limitation is known as "variable scope".

① Start a new program by specifying the C++ library classes to include, and a namespace prefix to use
```
#include <iostream>
using namespace std ;
```

scope.cpp

② Next, declare two simple function prototypes
```
float bodyTempC() ;
float bodyTempF() ;
```

③ Now, add a main function containing calls to each function and a final **return** statement
```
int main()
{
  cout << "Centigrade: " << bodyTempC() << endl ;
  cout << "Fahrenheit: " << bodyTempF() << endl ;
  return 0 ;
}
```

④ After the main function, define both other functions – to each return the value of a local "temperature" variable
```
float bodyTempC()
{
  float temperature = 37.0 ;
  return temperature ;
}

float bodyTempF()
{
  float temperature = 98.6 ;
  return temperature ;
}
```

Variables of the same name do not conflict when they are declared in a different scope – they are not visible to each other.

⑤ Save, compile, and run the program to see the output

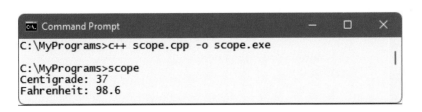

Passing arguments

Function calls frequently supply argument values to a function. These can be of any quantity and data type, but they must agree with those specified in the function prototype declaration.

Note that arguments passed to a function only supply a copy of the original value, in a procedure known as "passing by value".

The values passed to arguments can be "static" values, specified in the program code, or "dynamic" values that are input by the user. At a command prompt, the C++ **cin** function can be used with the **>>** input stream operator to direct a value from standard input to a variable, like this:

```
float num ;
cout << "Please enter a number: " ;
cin >> num ;
```

Input can then be passed to a function as an argument in a function call, such as **workWith(num)**.

Optionally, a function prototype can assign default values to arguments, which will be used when a call does not pass an argument value. Multiple arguments can be assigned default values in the prototype but these must always appear at the end of the argument list, after any other arguments.

In the same way that functions can be called from the main function, functions may call other functions and pass arguments.

Beware

Function prototypes must be declared before they can be defined. Typically, the prototypes appear before the main function, and their definitions appear after the main function.

C++

args.cpp

1. Start a new program by specifying the C++ library classes to include, and a namespace prefix to use
```
#include <iostream>
using namespace std ;
```

2. Next, declare a function prototype that returns a float value and specifies a single float argument, to which a default value is assigned
```
float fToC ( float degreesF = 32.0 ) ;
```

3. Add a main function containing a final **return** statement
```
int main()
{
   // Program code goes here.
   return 0 ;
}
```

4 After the main function, define the "fToC" function with statements that will return a converted value

```
float fToC( float degreesF )
{
  float degreesC = ( ( 5.0 / 9.0 ) * ( degreesF - 32.0 ) ) ;
  return degreesC ;
}
```

5 In the main function, insert a statement to declare two **float** variables – to store an input Fahrenheit temperature value and its Centigrade equivalent

```
float fahrenheit, centigrade ;
```

6 Insert statements to request that user input be stored in the first variable

```
cout << "Enter a Fahrenheit temperature:\t" ;
cin >> fahrenheit ;
```

7 Next, call the "fToC" function to convert the input value – and assign the conversion to the second variable

```
centigrade = fToC( fahrenheit ) ;
```

8 Now, output a message describing the result

```
cout << fahrenheit << "F is " << centigrade << "C" ;
```

9 Finally, add a statement to output a further message using the default argument value of the function prototype

```
cout << endl << "Freezing point: " << fToC() << "C" ;
```

10 Save, compile, and run the program, then enter a numeric value when requested to see the output

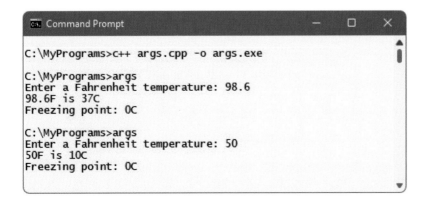

Hot tip

In the same way that functions can be called from the main function, functions may call other functions and pass arguments to them.

Don't forget

The names given to the arguments and variables in the function definition do not need to be the same as the variable names in the calling function – but it helps to clarify the program.

Overloading functions

Function "overloading" allows functions of the same name to happily co-exist in the same program, providing their arguments differ in number, data type, or both number and data type. The compiler matches a function call to the correct version of the function by recognizing its argument number and data types – a process known as "function resolution".

It is useful to create overloaded functions when the tasks they are to perform are similar, yet subtly different.

overload.cpp

1   Start a new program by specifying the C++ library classes to include, and a namespace prefix to use
```
#include <iostream>
using namespace std ;
```

2   Below the preprocessor instructions, declare a function prototype that returns a float value and has one argument
```
float computeArea ( float ) ;
```

3   Now, declare two overloaded function prototypes – having different arguments to the first prototype
```
float computeArea ( float, float ) ;
float computeArea ( char, float, float ) ;
```

4   Below the prototype declarations, add a main function containing a final **return** statement
```
int main()
{
  // Program code goes here.
  return 0 ;
}
```

5   After the main function, define the first function that receives just one argument
```
float computeArea( float diameter )
{
  float radius = ( diameter / 2 ) ;
  return ( 3.141593 * ( radius * radius ) ) ;
}
```

Beware

Functions that only differ by their return data type cannot be overloaded – it's the arguments that must differ. Function resolution does not take the return data types into consideration.

6 Below the first function definition, define the overloaded functions that receive different arguments

```
float computeArea( float width, float height )
{
  return ( width * height ) ;
}

float computeArea( char letter, float width , float height )
{
  return ( ( width / 2 ) * height ) ;
}
```

7 In the main function, insert statements to declare two variables, and initialize one with user input

```
float num, area ;

cout << "Enter dimension in feet: " ;
cin >> num ;
```

8 Call the first function and output its returned value

```
area = computeArea( num ) ;
cout << "Circle: Area = " << area << " sq.ft." << endl ;
```

9 Call the overloaded functions and output their returns

```
area = computeArea( num, num ) ;
cout << "Square: Area = "<< area << " sq.ft." << endl ;
area = computeArea( 'T', num, num ) ;
cout << "Triangle: Area = "<< area << "sq.ft." << endl ;
```

10 Save, compile, and run the program, then enter a numeric value when requested, to see the output

Don't forget

The value passed to the **char** argument is never used – that argument is included merely to differentiate that overloaded function.

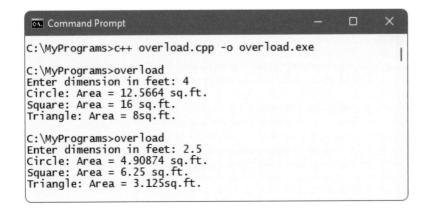

```
Command Prompt                              —    □    ✕

C:\MyPrograms>c++ overload.cpp -o overload.exe

C:\MyPrograms>overload
Enter dimension in feet: 4
Circle: Area = 12.5664 sq.ft.
Square: Area = 16 sq.ft.
Triangle: Area = 8sq.ft.

C:\MyPrograms>overload
Enter dimension in feet: 2.5
Circle: Area = 4.90874 sq.ft.
Square: Area = 6.25 sq.ft.
Triangle: Area = 3.125sq.ft.
```

Optimizing functions

Functions can call themselves recursively, to repeatedly execute the statements contained in their function body – much like a loop. As with loops, a recursive function must contain an incrementer and a conditional test to call itself again, or stop repeating when a condition is met. The syntax of a recursive function looks like this:

return-data-type function-name (argument-list)
{
       statements-to-be-executed ;
       incrementer ;
       conditional-test-to-recall-or-exit ;
}

The incrementer will change the value of a passed argument – so subsequent calls will pass the adjusted value back to the function.

optimize.cpp

1   Start a new program by specifying the C++ library classes to include, and a namespace prefix to use
```
#include <iostream>
using namespace std ;
```

2   Below the preprocessor instructions, declare two function prototypes that will both be recursive functions
```
int computeFactorials ( int, int ) ;
int factorial ( int ) ;
```

Hot tip

A recursive function generally uses more system resources than a loop – but it can make for more readable code.

3   Below the prototype declarations, add a main function containing a final **return** statement
```
int main()
{
  // Program code goes here.
  return 0 ;
}
```

4   After the main function, add the definition for the first function prototype – a recursive function
```
int computeFactorials( int num, int max )
{
  cout << "Factorial of " << num << ": " ;
  cout << factorial( num ) << endl ;     // Statements.
  num++ ;                                // Incrementer.
  if ( num > max ) return 0 ;      // Exit or call again.
  else return computeFactorials( num , max ) ;
}
```

5 Define a recursive function for the second prototype

```
int factorial( int n )
{
  int result ;
  if ( n == 1 ) result = 1 ;           // Exit or...
  else result = ( factorial( n - 1 ) * n ) ;   // Decrement...
  return result ;                      // and call again.
}
```

If you accidentally run an infinite recursive function, press the **Ctrl** + **C** keys to terminate the process.

6 At the start of the main function, insert a call to the recursive function

```
computeFactorials( 1, 8 ) ;
```

7 Save, compile, and run the program to see the output

```
Command Prompt                                   —   □   ×

C:\MyPrograms>c++ optimize.cpp -o optimize.exe

C:\MyPrograms>optimize
Factorial of 1: 1
Factorial of 2: 2
Factorial of 3: 6
Factorial of 4: 24
Factorial of 5: 120
Factorial of 6: 720
Factorial of 7: 5040
Factorial of 8: 40320

C:\MyPrograms>
```

The output lists factorial values (factorial 3 is 3x2x1=6, etc.), but the program can be improved by optimizing the **factorial()** function. This function does not need a variable if written with the ternary operator. It then contains just one statement, so its definition can replace the prototype declaration as an "inline" declaration. This means that the program need not keep checking between the declaration and definition, and so improves efficiency.

Inline declarations may only contain one or two statements, as the compiler recreates them at each calling point – longer inline declarations would, therefore, produce a more unwieldy program.

8 Delete the **factorial()** function definition, then replace its prototype declaration with this inline declaration

```
inline int factorial( int n )
{ return ( n == 1 ) ? 1 : ( factorial( n - 1 ) * n ) ; }
```

9 Save, compile, and run the program again to see the same output, produced more efficiently

Summary

- An **if** statement evaluates a given test expression for a Boolean value of **true** or **false**.

- Statements contained in braces after an **if** statement will only be executed when the evaluation is found to be **true**.

- The **if** and **else** keywords are used to perform conditional branching according to the result of a tested expression.

- A **switch** statement is an alternative form of conditional branching that matches a **case** statement to a given value.

- The **for** loop structure has parameters declaring an initializer, a test expression, and an incrementer or decrementer.

- A **while** loop and **do-while** loop must always have an incrementer or decrementer within their loop body.

- Any type of loop can be immediately terminated by including a **break** statement within the loop body.

- A single iteration of any type of loop can be skipped by including a **continue** statement within the loop body.

- Functions are usually declared as prototypes at the start of the program, and defined after the main function.

- Variables declared in a function are only accessible from within that function, as they only have local scope.

- Values can be passed into functions if arguments are declared in the function prototype and definition.

- Overloaded functions have the same name but a different number or type of declared arguments.

- Recursive functions repeatedly call themselves until a test condition is met.

- Short function definitions of just one or two statements can be declared in place of a prototype using the **inline** keyword.

4 Handling strings

This chapter demonstrates how to manipulate C++ text strings as a simpler, more powerful alternative to character arrays.

Creating string variables

Unlike the **char**, **int**, **float**, **double**, and **bool** data types, there is no native "string" data type in C++ – but its **<string>** library class provides a **string** object that emulates a string data type. To make this available to a program, the library must be added with an **#include <string>** directive at the start of the program.

Like the **<iostream>** class library, the **<string>** library is part of the **std** namespace that is used by the C++ standard library classes. This means that a string object can be referred to as **std::string**, or more simply as **string** when a **using namespace std**; directive is included at the start of the program.

Once the **<string>** library is made available, a **string** "variable" can be declared in the same way as other variables. The declaration may optionally initialize the variable using the = assignment operator, or it may be initialized later in the program.

Additionally, a **string** variable may be initialized by including a text string between parentheses after the variable name.

Text strings in C++ must always be enclosed within " " double quote characters – ' ' single quotes are only used to surround character values of the **char** data type.

A C++ **string** variable is much easier to work with than the **char** arrays that C programmers must use, as it automatically resizes to accommodate the length of any text string. At a lower level, the text is still stored as a character array, but the **string** variable lets you ignore those details. Consequently, a character array can be assigned to a **string** variable using the = assignment operator.

It is important to remember that when numeric values are assigned to a **string** variable, they are no longer a numeric data type, so arithmetic cannot be performed on them. For example, attempting to add **string** values of "7" and "6" with the + addition operator produces the concatenated **string** "76", not the numerical value of 13. In this case, the + operator recognizes that the context is not arithmetical, so adopts the guise of "concatenation operator" to unite the two strings. Similarly, the += operator appends a **string** to another **string** and is useful to build long strings of text.

Several **string** values are built into a single long string in the example program described on the opposite page.

1 Start a new program by specifying the C++ library classes to include, and a namespace prefix to use

```cpp
#include <string>            // Include string support.
#include <iostream>
using namespace std ;
```

string.cpp

2 Add a main function containing four **string** variable declarations and a final **return** statement

```cpp
int main()
{
  string text = "9" ;
  string term( "9 " ) ;
  string info = "Toys" ;
  string color ;
  // Add more statements here.
  return 0 ;
}
```

3 In the main function, after the variable declarations, insert statements to declare and initialize a character array, then assign its value to the uninitialized **string** variable

```cpp
char hue[4] = { 'R', 'e', 'd', '\0' } ;
color = hue ;
```

4 Assign a longer text **string** to one of the **string** variables

```cpp
info = "Balloons" ;
```

5 Build a long **string** by combining all the **string** variable values in the first **string** variable, then output the combined **string** value

```cpp
text += ( term + color + info ) ;
cout << endl << text << endl ;
```

Remember to add the special **\0** character to mark the end of a **string** in a **char** array.

6 Save, compile, and run the program to see the output

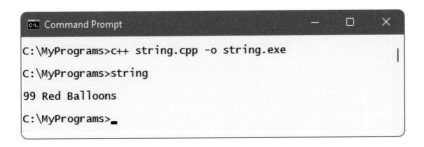

63

Getting string input

The C++ **cin** function, which was introduced in Chapter 3 to input numeric values, can also assign text input to **string** variables. This has a limitation, as it can only be used to input a single word at a time – the **cin** function stops reading the input when it encounters a space, leaving any other text in the "input buffer".

When you want to allow the user to input a string with spaces, such as a sentence, the **getline()** function can be used. This function requires two arguments to specify the source and destination of the string. For example, where the **cin** function is the source, and a string variable named "str" is the destination:

getline(cin , str) ;

The **getline()** function reads from an input "stream" until it encounters a **\n** newline character at the end of the line – created when you hit **Return**.

Care must be taken when mixing **cin** and **getline()** functions, as the **getline()** function will automatically read anything left on the input buffer – giving the impression that the program is skipping an instruction. The **cin.ignore()** function can be used to overcome this problem by ignoring content left in the input buffer.

input.cpp

1 Start a new program by specifying the C++ library classes to include, and a namespace prefix to use
```
#include <string>          // Include string support.
#include <iostream>
using namespace std ;
```

2 Add a main function containing one string variable declaration and a final **return** statement
```
int main()
{
  string name ;
  // Add more statements here.
  return 0 ;
}
```

3 In the main function, insert statements assigning the user name input to a string variable, then outputting its value
```
cout << "Please enter your full name: " ;
cin >> name ;
cout << "Welcome " << name << endl ;
```

...cont'd

4 Next, insert a statement requesting the user name again, but this time assigning the input to the **string** variable with the **getline** function before outputting its value
```
cout << "Please re-enter your full name: " ;
getline( cin , name ) ;
cout << "Thanks, " << name << endl ;
```

Use the **cin** function for numeric input or single word input, but use the **getline()** function for string input.

5 Save, compile, and run the program and enter your full name when requested

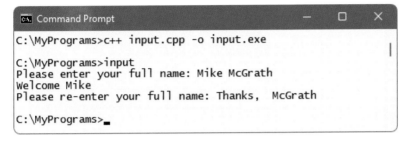

```
Command Prompt                                    —   □   ×

C:\MyPrograms>c++ input.cpp -o input.exe

C:\MyPrograms>input
Please enter your full name: Mike McGrath
Welcome Mike
Please re-enter your full name: Thanks,  McGrath

C:\MyPrograms>
```

This unsatisfactory result shows that **cin** reads up to the first space, leaving the second name in the input buffer, which is then read by **getline()** and subsequently output. The problem persists even when you enter only your first name, because **cin** leaves the newline character, created when you hit **Return**, on the input buffer.

The arguments to the **cin.ignore()** function specify it should discard up to 256 characters and stop when it encounters a newline character.

6 Edit the program to resolve this issue by inserting a statement, just before the call to the **getline** function, instructing it to ignore content in the input buffer
```
cin.ignore( 256, '\n' ) ;
```

7 Save, compile, and run the program again, then re-enter your full name to see the program perform as required

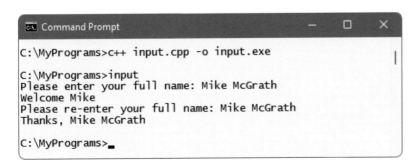

```
Command Prompt                                    —   □   ×

C:\MyPrograms>c++ input.cpp -o input.exe

C:\MyPrograms>input
Please enter your full name: Mike McGrath
Welcome Mike
Please re-enter your full name: Mike McGrath
Thanks, Mike McGrath

C:\MyPrograms>
```

Solving the string problem

A problem arises with **string** variables when you need to convert them to a different data type, perhaps to perform arithmetical operations with those values. As the **string** object is not a native C++ data type, a **string** variable value cannot be converted to an **int** or any other regular data type by casting.

The solution is provided by the C++ **<sstream>** library that allows a **stringstream** object to act as an intermediary, through which **string** values can be converted to a numeric data type, and numeric values can be converted to a **string** data type. To make this ability available to a program, the library must be added with an **#include <sstream>** directive at the start of the program.

Values can be loaded into a **stringstream** object with the familiar output stream **<<** operator that is used with **cout** statements. Contents can then be extracted from a **stringstream** object with the **>>** input stream operator that is used with **cin** statements.

In order to re-use a **stringstream** object, it must first be returned to its original state. This requires its contents to be set as an empty **string** and its status bit flags to be cleared.

convert.cpp

1 Start a new program by specifying the C++ library classes to include, and a namespace prefix to use

```
#include <string>        // Include string support.
#include <sstream>       // Include stringstream support.
#include <iostream>
using namespace std ;
```

2 Add a main function containing a final **return** statement and declaring two initialized variables to be converted

```
int main()
{
  string term = "100" ;
  int number = 100 ;
  // Add more statements here.
  return 0 ;
}
```

3 In the main function, insert statements to declare an integer variable, **string** variable, and a **stringstream** object

```
int num ;                // To store a converted string.
string text ;            // To store a converted integer.
stringstream stream ;    // To perform conversions.
```

4 Next, use the stream output operator to load the
initialized **string** value into the **stringstream** object
stream << term ;   // Load the string.

5 Use the stream input operator to extract content from the
stringstream object into the uninitialized integer variable
stream >> num ;   // Extract the integer.

6 Perform arithmetic on the integer and output the result
num /= 4 ;
cout << "Integer value: " << num << endl ;

7 Reset the **stringstream** object ready for re-use
stream.str("") ;   // Empty the contents.
stream.clear() ;   // Empty the bit flags.

8 Now, use the stream output operator to load the
initialized integer value into the **stringstream** object
stream << number ;   // Load the integer.

9 Use the stream input operator to extract content from the
stringstream object into the uninitialized **string** variable
stream >> text ;   // Extract the string.

10 Perform concatenation on the **string** and output the result
text += " Per Cent" ;
cout << "String value: " << text << endl ;

11 Save, compile, and run the program to see the converted
output values

Hot tip

Notice how the
stringstream object's
str() function is used
here to reset its contents
to an empty string.

Don't forget

A non-empty
stringstream object
has bit flags indicating
its status as **good**,
bad, **eof**, or **fail** –
these should be cleared
before re-use by the
stringstream object's
clear() function, as
demonstrated here.

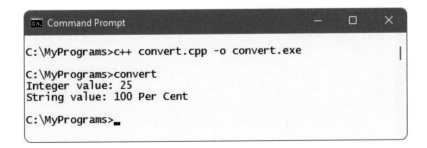

```
Command Prompt                                    —    □    ×

C:\MyPrograms>c++ convert.cpp -o convert.exe            |

C:\MyPrograms>convert
Integer value: 25
String value: 100 Per Cent

C:\MyPrograms>_
```

Discovering string features

The C++ **<string>** library provides a number of functions that make it easy to work with strings. To use them, simply add the function name after the **string** variable name and a dot. For example, with a **stringstream** variable named "msg" you can call upon the **size()** function, to return its character length, with **msg.size()**.

A **string** variable can be emptied of all characters by assigning it an empty string with two double quotes without spacing – as "", or alternatively by calling the **<string>** library's **clear()** function.

Unlike a **char** array, a **string** variable will dynamically enlarge to accommodate the number of characters assigned to it, and its current memory size can be revealed with the **<string>** library's **capacity()** function. Once enlarged, the allocated memory size remains, even when a smaller **string** gets assigned to the variable.

The **<string>** library's **empty()** function returns a Boolean **true** (1) or **false** (0) response to reveal whether the string is empty or not.

features.cpp

1 Start a new program by specifying the C++ library classes to include, and a namespace prefix to use

```
#include <string>            // Include string support.
#include <iostream>
using namespace std ;
```

2 Below the preprocessor directives, declare a function prototype with a single **string** data type argument

```
void computeFeatures( string ) ;
```

3 Add a main function containing a final **return** statement and declaring an initialized **string** variable

```
int main()
{
  string text = "C++ is fun" ;
  // Add more statements here.
  return 0 ;
}
```

The **length()** function can be used in place of the **size()** function to reveal the size of a string value.

4 After the main function, define the declared function to display the **string** variable value when called

```
void computeFeatures( string text )
{
  cout << endl << "String: " << text << endl ;
}
```

5 In the function definition, add statements to output features of the **string** variable
```
cout << "Size:   " << text.size() ;
cout << "    Capacity: " << text.capacity() ;
cout << "    Empty?: " << text.empty() << endl ;
```

6 In the main function, insert a call to the defined function
```
computeFeatures( text ) ;
```

7 Next, in the main function, insert a statement to enlarge the **string** value and call the function to see its features
```
text += " for everyone" ;
computeFeatures( text ) ;
```

8 Now, insert a statement to reduce the **string** value
```
text = "C++ Fun" ;
computeFeatures( text ) ;
```

9 Finally, insert a statement to empty the **string** variable
```
text.clear() ;
computeFeatures( text ) ;
```

10 Save, compile, and run the program to see the output

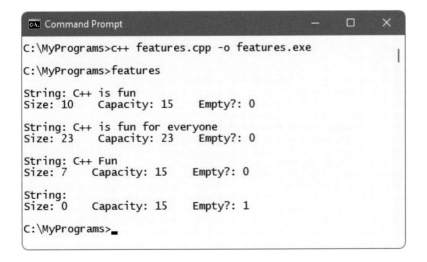

```
Command Prompt                                    —    □    ×

C:\MyPrograms>c++ features.cpp -o features.exe

C:\MyPrograms>features

String: C++ is fun
Size: 10    Capacity: 15    Empty?: 0

String: C++ is fun for everyone
Size: 23    Capacity: 23    Empty?: 0

String: C++ Fun
Size: 7    Capacity: 15    Empty?: 0

String:
Size: 0    Capacity: 15    Empty?: 1

C:\MyPrograms>_
```

The **empty()** function is useful to check if the user has entered requested input.

A space occupies one memory element – just like a character does.

69

Joining & comparing strings

When the **+** operator is used to concatenate strings in an assignment, the combined strings get stored in the **string** variable. But when it is used with the **cout** function, the strings are only combined in the output – the variable values are unchanged.

The **<string>** library's **append()** function can also be used to concatenate strings, specifying the **string** value to append as an argument within its parentheses. When this is used with the **cout** function, the strings are combined in the variable, then its value written as output – in this case, the variable value does change.

String comparisons can be made, in the same way as numeric comparisons, with the **==** equality operator. This returns **true** (1) when both strings precisely match, otherwise it returns **false** (0).

Alternatively, the **<string>** library's **compare()** function can be used to compare a **string** value specified as its argument. Unlike the **==** equality comparison, the **compare()** function returns 0 when the strings are identical, by examining the **string** value's combined ASCII code values. When the **string** argument totals more than the first string, it returns -1, otherwise it returns 1.

compare.cpp

1 Start a new program by specifying the C++ library classes to include, and a namespace prefix to use
```
#include <string>              // Include string support.
#include <iostream>
using namespace std ;
```

2 Add a main function containing a final **return** statement and declaring three initialized **string** variables
```
int main()
{
  string lang = "C++" ;
  string term = " Programming" ;
  string text = "C++ Programming" ;
  // Add more statements here.
  return 0 ;
}
```

3 In the main function, insert statements to output two **string** values combined with the + concatenate operator and the (unchanged) value of the first variable
```
cout << "Concatenated: " << ( lang + term ) << endl ;
cout << "Original: " << lang << endl ;
```

Don't forget

The **+=** assignment operator can also be used to append a string.

4 Next, insert statements to output two **string** values combined with the **append()** function and the (changed) value of the first variable
```
cout << "Appended: " << lang.append( term ) << endl ;
cout << "Original: " << lang << endl << endl ;
```

5 Use the **==** equality operator to compare two **string** values that differ, then two **string** values that match
```
cout << "Differ: " << ( lang == term ) << endl ;
cout << "Match: " << ( lang == text ) << endl << endl ;
```

6 Now, use the **compare()** function to compare three **string** values, examining their ASCII code total values
```
cout << "Match: " << lang.compare( text ) << endl ;
cout << "Differ: " << lang.compare( term ) << endl ;
cout << "Lower ASCII: " <<
                lang.compare( "zzzzz" ) << endl ;
```

7 Save, compile, and run the program to see the output

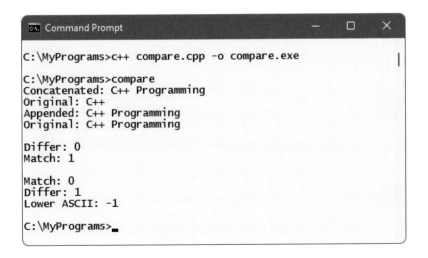

```
Command Prompt                                    —  □  ✕

C:\MyPrograms>c++ compare.cpp -o compare.exe    |

C:\MyPrograms>compare
Concatenated: C++ Programming
Original: C++
Appended: C++ Programming
Original: C++ Programming

Differ: 0
Match: 1

Match: 0
Differ: 1
Lower ASCII: -1

C:\MyPrograms>_
```

Hot tip

In comparisons, character order is taken into account – so comparing "za" to "az" reveals that "za" has a greater total. In terms of ASCII values, 'a' is 97, and 'z' is 122.

71

Copying & swapping strings

String values can be assigned to a **string** variable by the = assignment operator, or by the **<string>** library's **assign()** function. This function specifies the **string** value to be copied to the variable as an argument within its parentheses.

Optionally, the **assign()** function can copy just a part of the specified **string** value by stating the position of the starting character as a second argument, and the number of characters to copy as a third argument.

The contents of a **string** variable can be exchanged for that of another **string** variable by the **<string>** library's **swap()** function. In this case, the contents of the first variable receives those of the second variable, which in turn receives those of the first variable.

swap.cpp

1 Start a new program by specifying the C++ library classes to include, and a namespace prefix to use
```
#include <string>              // Include string support.
#include <iostream>
using namespace std ;
```

2 Add a main function containing a final **return** statement and declaring three **string** variables – with one initialized
```
int main()
{
  string front ;
  string back ;
  string text =
    "Always laugh when you can. It\'s cheap medicine." ;
  // Add more statements here.
  return 0 ;
}
```

3 In the main function, insert a statement to assign the entire value of the initialized **string** variable to the first uninitialized **string** variable
```
front.assign( text ) ;
```

4 Next, insert a statement to output the newly assigned **string** value
```
cout << endl << "Front: " << front << endl ;
```

5 Now, insert a statement to assign only the first 27 characters of the initialized variable to the first variable
front.assign(text, 0, 27) ;

Hot tip

Use the **=** assignment operator to assign complete strings, and the **assign()** function to assign partial strings.

6 Output the newly assigned **string** value
cout << endl << "Front: " << front << endl ;

7 Next, assign only the last part of the initialized **string** variable to the second uninitialized variable, starting at character (element) 27
back.assign (text, 27 , text.size()) ;

8 Now, output this newly assigned **string** value
cout << "Back: " << back << endl ;

9 Finally, exchange the assigned **string** values contained in the first and second **string** variables, then output the exchanged values
back.swap(front) ;
cout << endl << "Front: " << front << endl ;
cout << "Back: " << back << endl ;

Hot tip

Use the **swap()** function wherever possible, rather than creating additional string variables.

10 Save, compile, and run the program to see the output

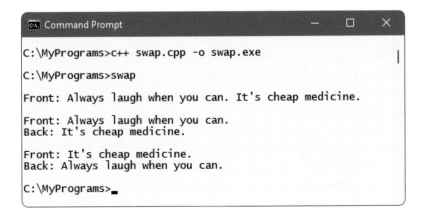

```
C:\MyPrograms>c++ swap.cpp -o swap.exe

C:\MyPrograms>swap

Front: Always laugh when you can. It's cheap medicine.

Front: Always laugh when you can.
Back: It's cheap medicine.

Front: It's cheap medicine.
Back: Always laugh when you can.

C:\MyPrograms>_
```

Finding substrings

A **string** value can be searched to see if it contains a specified "substring" using the **find()** function of the **<string>** library. Its parentheses should specify the substring to seek as its first argument, and the index number of the character at which to start searching as its second argument.

When a search successfully locates the specified substring, the **find()** function returns the index number of the first occurrence of the substring's first character within the searched string. When the search fails, **find()** returns a value of -1 to indicate failure.

There are several other functions in the **<string>** library that are related to the **find()** function. Two of these are the **find_first_of()** function and the **find_first_not_of()** function. Instead of seeking the first occurrence of a complete **string**, as **find()** does, the **find_first_of()** function seeks the first occurrence of any of the characters in a specified **string**, and **find_first_not_of()** seeks the first occurrence of a character that is not in the specified **string**.

The **find_last_of()** and **find_last_not_of()** functions work in a similar manner – but begin searching at the end of the **string** then move forward.

find.cpp

1   Start a new program by specifying the C++ library classes to include, and a namespace prefix to use
```
#include <string>              // Include string support.
#include <iostream>
using namespace std ;
```

2   Add a main function containing a final **return** statement, an initialized string variable declaration, and declaring an integer variable to store search results
```
int main()
{
  string text = "I can resist anything but temptation." ;
  int num ;
  // Add more statements here.
  return 0 ;
}
```

3   In the main function, insert statements to output the start position of a substring within the entire **string** variable
```
num = text.find( "resist", 0 ) ;
cout << "Position: " << num << endl ;
```

…cont'd

4 Next, insert a statement to seek a non-existent substring within the entire **string** variable and output the result
```
num = text.find( "nonsuch" , 0 ) ;
cout << "Result:  " << num << endl ;
```

Beware

The searches are case sensitive, so seeking "If" and "if" may produce different results – here, uppercase 'I' matches.

5 Now, insert a statement to output the start position of the first occurrence of any characters in an "If" substring found within the entire **string** variable
```
num = text.find_first_of( "If" ) ;
cout << "First I: " << num << endl ;
```

6 Insert a statement to report the string position of the first character not within the "If" substring
```
num = text.find_first_not_of( "If" ) ;
cout << "First not I: " << num << endl ;
```

7 Next, insert a statement to seek the last occurrence of the letter "t" within the **string** variable and output its position
```
num = text.find_last_of( "t" ) ;
cout << "Last t: " << num << endl ;
```

Don't forget

The first character in a string is at position 0, not at position 1.

8 Now, add a statement to report the **string** position of the last character within the **string** variable that is not a "t"
```
num = text.find_last_not_of( "t" ) ;
cout << "Last not t: " << num << endl ;
```

9 Save, compile, and run the program to see the search results indicating failure or the positions when located

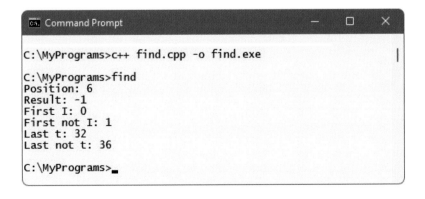

75

Replacing substrings

The **<string>** library contains a number of useful functions to manipulate substrings. A **string** can be inserted into another **string** using the **insert()** function. This requires the index position at which the **string** should be inserted as its first argument, and the **string** value to be inserted as its second argument.

Conversely, a substring can be removed from a **string** using the **erase()** function. This requires the index position at which it should begin erasing as its first argument, and the number of characters to be erased as its second argument.

The **replace()** function neatly combines the **erase()** and **insert()** functions to both remove a substring and insert a replacement. It requires three arguments specifying the position at which it should begin erasing, the number of characters to be erased, and the replacement **string** to be inserted at that position.

A substring can be copied from a **string** using the **substr()** function, stating the index position at which it should begin copying as its first argument, and the number of characters to be copied as its second argument.

The character at any specified position within a **string** can be copied using the **at()** function, which requires the index position as its argument. The final character in a **string** always has an element index number one less than the length of the **string** – because index numbering starts at 0, not 1.

sub.cpp

① Start a new program by specifying the C++ library classes to include, and a namespace prefix to use
```cpp
#include <string>          // Include string support.
#include <iostream>
using namespace std ;
```

② Add a main function containing a final **return** statement, an initialized **string** variable declaration, and a statement outputting the **string** variable value
```cpp
int main()
{
  string text = "I do like the seaside" ;
  cout << "Original: " << text << endl ;
  // Add more statements here.
  return 0 ;
}
```

3 In the main function, insert statements to insert a substring into the variable value at index position 10, and to output the modified **string**
```
text.insert( 10, "to be beside " ) ;
cout << "Inserted: " << text << endl ;
```

4 Next, insert statements to erase two characters from the modified **string** value starting at index position 3, and to output the revised **string**
```
text.erase( 2, 3 ) ;
cout << "Erased:  " << text << endl ;
```

5 Now, insert statements to remove 25 characters at index position 7, insert a replacement substring, then output the revised **string** again
```
text.replace( 7, 25, "strolling by the sea" ) ;
cout << "Replaced:  " << text << endl ;
```

6 Finally, insert statements to output nine copied characters at index position 7, and to output the final character in the string
```
cout << "Copied: " << text.substr( 7, 9 ) << endl ;
cout << "Last character: " <<
              text.at( text.size() - 1 ) << endl ;
```

7 Save, compile, and run the program to see the output showing how the **string** has been manipulated

The **insert()** function can optionally have a third and fourth argument – specifying the position in the substring at which to begin copying, and the number of characters to be copied.

The **replace()** function can optionally have a fourth and fifth argument – specifying the position in the substring at which to begin copying, and the number of characters to be copied.

77

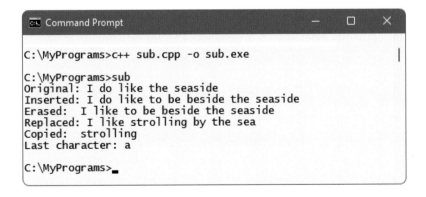

```
Command Prompt                                  ─   □   ✕

C:\MyPrograms>c++ sub.cpp -o sub.exe            |

C:\MyPrograms>sub
Original: I do like the seaside
Inserted: I do like to be beside the seaside
Erased:  I like to be beside the seaside
Replaced: I like strolling by the sea
Copied:  strolling
Last character: a

C:\MyPrograms>_
```

Summary

- The C++ **<string>** library provides a "string" object that emulates a data type – so that **string** variables can be created.

- Arithmetic cannot be performed on numeric values assigned to **string** variables until they are converted to a numeric data type.

- The standard **cin** function reads from standard input until it encounters a space, so can be used to input a single word, and provides an **ignore()** function to disregard the input buffer.

- The **getline()** function reads from standard input until it encounters a newline, so can be used to input a string of text.

- The C++ **<sstream>** library provides a "stringstream" object that acts an intermediary to convert strings to other data types.

- A **string** variable can be emptied by assigning it an empty string (= "") or by calling its **clear()** function.

- Features of a **string** variable can be revealed by calling its **size()**, **capacity()**, and **empty()** functions.

- Multiple **string** values can be concatenated by the + operator.

- A **string** can be appended to another **string** by the += operator or by calling its **append()** function.

- A **string** can be compared to another **string** by the == operator or by calling its **compare()** function.

- A **string** value can be assigned to a **string** variable using the = operator or by calling its **assign()** function.

- The **swap()** function swaps the values of two **string** variables.

- Substrings of a **string** can be sought with the **find()** function, or specialized functions such as **find_first_of()**, and a character retrieved from a specified index position by the **at()** function.

- A substring can be added to a **string** by its **insert()** function, removed by its **erase()** function, replaced by its **replace()** function, or copied by its **substr()** function.

5 Reading and writing files

This chapter demonstrates how to store and retrieve data in text files, and illustrates how to avoid errors in C++ programs.

Writing a file

The ability to read and write files from a program provides a useful method of maintaining data on the computer's hard disk. The format of the data may be specified as human-readable plain text format or machine-readable binary format.

The standard C++ **<fstream>** library provides functions for working with files, which can be made available by adding an **#include <fstream>** directive at the start of the program.

For each file that is to be opened, a filestream object must first be created. This will be either an "ofstream" (output filestream) object, for writing data to the file, or an "ifstream" (input filestream) object, for reading data from the file. The **ofstream** object is used like the **cout** function that writes to standard output, and the **ifstream** object works like the **cin** function that reads from standard input.

The declaration of a filestream object for writing output begins with the **ofstream** keyword, then a chosen name for that particular filestream object followed by parentheses nominating the text file to write to. So, the declaration syntax looks like this:

ofstream *object-name* (*"file-name"*) ;

The argument nominating the text file may optionally contain the full file path, such as **"C:\data\log.txt"** or **"/home/user/log.txt"**, otherwise the program will seek the file within the directory in which the program resides.

Before writing output to a file, the program should always first test that the filestream object has actually been created. Typically, this is performed by an **if** statement that allows the program to write output only when the test is successful.

If a nominated file already exists, it will by default be overwritten without warning. Otherwise, a new file will be created and written.

After writing output to a nominated file, the program should always call the associated filestream object's **close()** function to close the output filestream.

The program described opposite first builds a string for writing as output. This is written to a nominated file when the filestream object has been successfully created, then the filestream is closed.

Beware

The nominated file name or path must be enclosed within double quotes, like a **string**.

1 Start a new program by specifying the C++ library classes to include, and a namespace prefix to use

```cpp
#include <fstream>      // Include filestream support.
#include <string>
#include <iostream>
using namespace std ;
```

write.cpp

2 Add a main function containing a final **return** statement and building a lengthy text string in a **string** variable

```cpp
int main()
{
  string poem = "\n\tI never saw a man who looked" ;
  poem.append( "\n\tWith such a wistful eye" ) ;
  poem.append( "\n\tUpon that little tent of blue" ) ;
  poem.append( "\n\tWhich prisoners call the sky" ) ;
  // Add more statements here.
  return 0 ;
}
```

Hot tip

String values can contain \n newline and \t tab escape sequences for formatting lines.

3 In the main function, create an output filestream object

```cpp
ofstream writer( "poem.txt" ) ;
```

81

4 Insert statements to write the **string** to a file or exit, then save, compile, and run the program to see the result

```cpp
if ( ! writer )
{
  cout << "Error opening file for output" << endl ;
  return -1 ;      // Signal an error then exit the program.
}
writer << poem << endl ;      // Write output.
writer.close() ;              // Close filestream.
```

Don't forget

Notice how the newline and tab characters are preserved in the text file.

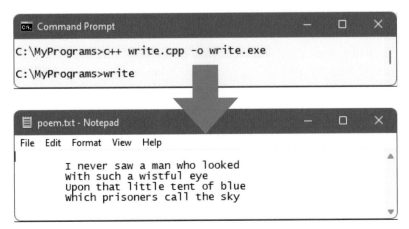

Appending to a file

When a filestream object is created, the parentheses following its chosen name can optionally contain additional arguments, specifying a range of file "modes" to control the behavior of that filestream object. These file modes are part of the **ios** namespace, so must be explicitly addressed using that prefix. Each file mode is listed in the table below, together with a behavior description:

Mode:	Behavior:
ios::out	Open a file to write output
ios::in	Open a file to read input
ios::app	Open a file to append output at the end of any existing content
ios::trunc	Truncate the existing file (default behavior)
ios::ate	Open a file without truncating and allow data to be written anywhere in the file
ios::binary	Treat the file as binary format rather than text so the data may be stored in non-text format

Don't forget

The preprocessor directive **using namespace std;** allows the **std** namespace prefix to be omitted – so **cout** refers to the **std::cout** function. The **ios** namespace exists within the **std** namespace – so the file modes can be explicitly addressed using both namespace prefixes – for example, **std::ios::out**

Multiple modes may be specified if they are separated by a "|" pipe character. For example, the syntax of a statement to open a file for binary output looks like this:

ofstream *object-name* (*"file-name"* , ios::out|ios::binary) ;

The default behavior when no modes are explicitly specified regards the file as a text file that will be truncated after writing.

The most commonly specified mode is **ios::app**, which ensures existing content will be appended, rather than overwritten, when new output is written to the nominated file.

The program described opposite appends data to the text file created in the previous example.

append.cpp

1 Start a new program by specifying the C++ library classes to include, and a namespace prefix to use

```cpp
#include <fstream>      // Include filestream support.
#include <string>
#include <iostream>
using namespace std ;
```

2 Add a main function containing a final **return** statement and building a text string in a **string** variable

```cpp
int main()
{
  string info = "\n\tThe Ballad of Reading Gaol" ;
  info.append( "\n\t\t\tOscar Wilde 1898" ) ;
  // Add more statements here.
  return 0 ;
}
```

3 In the main function, create an output filestream object – specifying a file mode that will append to existing text

```cpp
ofstream writer( "poem.txt" , ios::app ) ;
```

4 Insert statements to append the **string** to a file or exit, then save, compile, and run the program to see the result

```cpp
if ( ! writer )
{
  cout << "Error opening file for output" << endl ;
  return -1 ;     // Signal an error then exit the program.
}
writer << info << endl ;      // Append output.
writer.close() ;              // Close filestream.
```

The file must allow the program suitable read and write permissions.

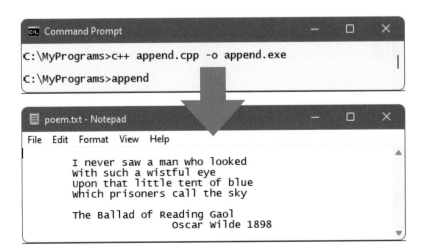

```
Command Prompt                              —   □   ×

C:\MyPrograms>c++ append.cpp -o append.exe

C:\MyPrograms>append
```

```
poem.txt - Notepad                          —   □   ×

File  Edit  Format  View  Help

        I never saw a man who looked
        with such a wistful eye
        Upon that little tent of blue
        which prisoners call the sky

        The Ballad of Reading Gaol
                        Oscar Wilde 1898
```

Reading characters & lines

The **ifstream** filestream object has a **get()** function that can be used in a loop to read a file and assign each character in turn to the **char** variable specified as its argument:

read.cpp

1 Start a new program by specifying the C++ library classes to include, and a namespace prefix to use
```
#include <fstream>      // Include filestream support.
#include <iostream>
using namespace std ;
```

2 Add a main function containing a final **return** statement and two variable declarations – one variable to store a character and another to count loop iterations
```
int main()
{
  char letter ;
  int i ;
  // Add more statements here.
  return 0 ;
}
```

3 In the main function, create an input filestream object to read the text file from the previous example
```
ifstream reader( "poem.txt" ) ;
```

4 Insert statements to exit unless the filestream object exists
```
if ( ! reader )
{
  cout << "Error opening input file" << endl ;
  return -1 ;      // Signal an error then exit the program.
}
```

Hot tip

Notice how the **ifstream eof()** function is used to check if the "end of file" has been reached.

5 Next, insert a loop to read the text file, assigning each character in turn to the variable and outputting its value
```
else
for ( i = 0 ; ! reader.eof() ; i++ )
{
  reader.get( letter ) ;
  cout << letter ;
}
```

6 Finally, insert statements to close the filestream and output the total number of loop iterations
```
reader.close() ;
cout << "Iterations: " << i << endl ;
```

7 Save, compile, and run the program to see the text file contents and loop count get displayed on standard output

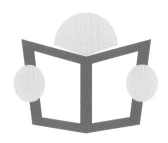

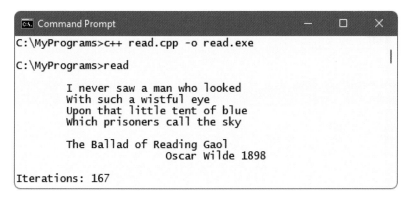

```
C:\MyPrograms>c++ read.cpp -o read.exe

C:\MyPrograms>read

        I never saw a man who looked
        With such a wistful eye
        Upon that little tent of blue
        Which prisoners call the sky

        The Ballad of Reading Gaol
                        Oscar Wilde 1898

Iterations: 167
```

This program works well enough but the loop must make many iterations to output the text file contents. Efficiency could be improved by reading a line on each iteration of the loop:

8 Insert a preprocessor directive to make the C++ string library available to the program
#include <string>

9 Replace the **char** variable declaration with a **string** variable declaration
string line ;

10 Replace both statements in the **for** loop to read lines, then save, compile, and run the program once more
getline(reader , line) ;
cout << line << endl ;

Output an **endl** after each line output – because **getline()** stops reading when it meets a **\n** newline character.

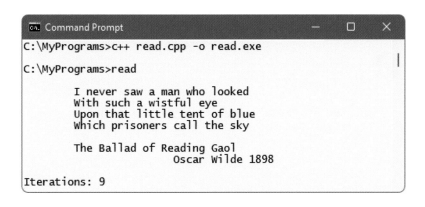

```
C:\MyPrograms>c++ read.cpp -o read.exe

C:\MyPrograms>read

        I never saw a man who looked
        With such a wistful eye
        Upon that little tent of blue
        Which prisoners call the sky

        The Ballad of Reading Gaol
                        Oscar Wilde 1898

Iterations: 9
```

Formatting with getline

The **getline()** function can optionally have a third argument to specify a delimiter at which to stop reading a line. This can be used to separate text read from a tabulated list in a data file:

format.cpp

1 In a plain text editor, create a text file containing 12 items of data of four items per line, each separated by a tab

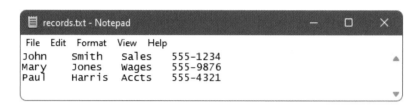

```
records.txt - Notepad
File   Edit   Format   View   Help
John       Smith      Sales      555-1234
Mary       Jones      Wages      555-9876
Paul       Harris     Accts      555-4321
```

2 Start a new program by specifying the C++ library classes to include, and a namespace prefix to use
```
#include <fstream>      // Include filestream support.
#include <string>
#include <iostream>
using namespace std ;
```

The **string** array must have a sufficient number of elements to store each item of data – it would need to be enlarged to handle more records.

3 Add a main function containing a final **return** statement and four variable declarations – a fixed number, a **string** array to store data, and two counter variables set to 0
```
int main()
{
  const int RANGE = 12 ;
  string tab[ RANGE ] ;
  int i = 0 , j = 0 ;
  // Add more statements here.
  return 0 ;
}
```

4 Insert a statement to create an input filestream object
```
ifstream reader( "records.txt" ) ;
```

5 Insert statements to exit unless the filestream object exists
```
if ( ! reader )
{
  cout << "Error opening input file" << endl ;
  return -1 ;
}
```

6 Next, insert a loop that will read each line into the **string** array – reading up to a **\t** tab for the first three items and up to a **\n** newline for the fourth item on each line

```
while ( ! reader.eof() )
{
  if ( ( i + 1 ) % 4 == 0 )
        getline( reader, tab[ i++ ], '\n' ) ;
  else
        getline( reader, tab[ i++ ], '\t' ) ;
}
```

7 Now, close the filestream and reset the counter

```
reader.close() ;
i = 0 ;
```

8 Insert a second loop to output the data stored in each array element, formatted with descriptions and newlines

```
while ( i < RANGE )
{
  cout << endl << "Record Number: " << ++j << endl ;
  cout << "Forename: " << tab[ i++ ] << endl ;
  cout << "Surname: "  << tab[ i++ ] << endl ;
  cout << "Department: " << tab[ i++ ] << endl ;
  cout << "Telephone: " << tab[ i++ ] << endl ;
}
```

9 Save, compile, and run the program to see the formatted output from the text file

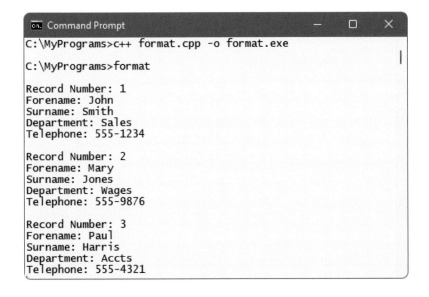

```
C:\MyPrograms>c++ format.cpp -o format.exe

C:\MyPrograms>format

Record Number: 1
Forename: John
Surname: Smith
Department: Sales
Telephone: 555-1234

Record Number: 2
Forename: Mary
Surname: Jones
Department: Wages
Telephone: 555-9876

Record Number: 3
Forename: Paul
Surname: Harris
Department: Accts
Telephone: 555-4321
```

Hot tip

The **if** statement tests if the item number (element number plus 1) is exactly divisible by four, to determine whether to read up to a newline or tab character.

Beware

The record counter must use a prefix incrementer to increase the variable value before it is output.

Manipulating input & output

The behavior of input and output streams can be modified using "insertion operators" with the **cout** and **cin** functions. Specifying an integer argument to their **width()** function sets the stream character width. Where the content does not fill the entire stream width, a fill character may be specified as the argument to their **fill()** function to indicate the empty portion. Similarly, the default precision of six decimal places for floating point numbers can be changed by specifying an integer to their **precision()** function. Statements using insertion operators to modify a stream should be made before those using the **<<** or **>>** operators.

The **<iostream>** library provides the "manipulators" listed in the table below, which modify a stream using the **<<** or **>>** operators.

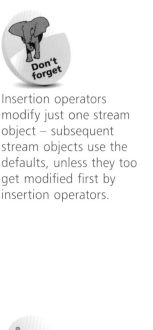

Don't forget

Insertion operators modify just one stream object – subsequent stream objects use the defaults, unless they too get modified first by insertion operators.

Hot tip

Manipulators marked with an * are the default behaviors.

Manipulator:	Display:
noboolalpha*	Boolean values as 1 or 0
boolalpha	Boolean values as "true" or "false"
dec*	Integers as base 10 (decimal)
hex	Integers as base 16 (hexadecimal)
oct	Integers as base 8 (octal)
right*	Text right-justified in the output width
left	Text left-justified in the output width
internal	Sign left-justified, number right-justified
noshowbase*	No prefix indicating numeric base
showbase	Prefix indicating numeric base
noshowpoint*	Whole number only when a fraction is 0
showpoint	Decimal point for all floating point numbers
noshowpos*	No + prefix before positive numbers
showpos	Prefix positive numbers with a + sign
noskipws*	Do not skip whitespace for >> input
skipws	Skip whitespace for >> input
fixed*	Floating point numbers to six decimal places
scientific	Floating point numbers in scientific notation
nouppercase*	Scientific as e and hexadecimal number as ff
uppercase	Scientific as E and hexadecimal number as FF

1 Start a new program by specifying the C++ library classes to include, and a namespace prefix to use

manipulate.cpp

```
#include <iostream>
using namespace std ;
```

2 Add a main function containing a final **return** statement and declaring two initialized variables

```
int main()
{
  bool isTrue = 1 ;
  int num = 255 ;
  // Add more statements here.
  return 0 ;
}
```

3 In the main function, insert statements to set the width and fill of an output stream, then output a text string on it

```
cout.width( 40 ) ;
cout.fill( '.' ) ;
cout << "Output" << endl ;
```

4 Next, insert statements to set the precision of an output stream to stop truncation of decimal places – then output a floating point number showing all its decimal places

```
cout.precision( 11 ) ;
cout << "Pi: " << 3.1415926536 << endl ;
```

Beware

Manipulators affect all input or output on that stream. For example, the **boolalpha** manipulator will display all Boolean values on that stream in written form.

5 Now, insert statements that use manipulators to output the variable values in modified formats

```
cout << isTrue << ": " << boolalpha << isTrue << endl ;
cout << num << ": " << hex << showbase
                    << uppercase << num << endl ;
```

6 Save, compile, and run the program to see the output

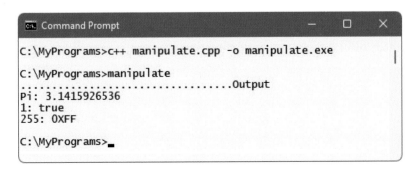

```
Command Prompt                              —   □   ×

C:\MyPrograms>c++ manipulate.cpp -o manipulate.exe

C:\MyPrograms>manipulate
...............................Output
Pi: 3.1415926536
1: true
255: 0XFF

C:\MyPrograms>_
```

Predicting problems

Despite the best efforts of the programmer, C++ programs may unfortunately contain one or more of these three types of bugs:

- **Syntax errors** – the code contains incorrect use of the C++ language. For example, an opening brace does not have a matching closing brace.

- **Logic errors** – the code is syntactically correct, but attempts to perform an operation that is illegal. For example, the program may attempt to divide a number by zero, causing an error.

- **Exception errors** – the program runs as expected until an exceptional condition is encountered that crashes the program. For example, the program may request a number, which the user enters in word form rather than in numeric form.

The C++ standards allow the compiler to spot "compile-time" errors involving syntax and logic, but the possibility of exceptional errors is more difficult to locate as they only occur at "run-time". This means that the programmer must try to predict problems that may arise and prepare to handle those exceptional errors.

The first step is to identify which part of the program code that may, under certain conditions, cause an exception error. This can then be surrounded by a "try" block, which uses the **try** keyword and encloses the suspect code within a pair of braces.

When an exception error occurs, the **try** block then "throws" the exception out to a "catch" block, which immediately follows the **try** block. This uses the **catch** keyword and encloses statements to handle the exception error within a pair of braces.

The program described on the opposite page has a **try** block containing a loop that increments an integer. When the integer reaches 5, a **throw()** function manually throws an exception to the **catch** block exception handler, passing the integer argument.

Hot tip

Always consider that the user will perform the unexpected – then ensure your programs can handle those actions.

1 Start a new program by specifying the C++ library classes to include, and a namespace prefix to use

```
#include <iostream>
using namespace std ;
```

try.cpp

2 Add a main function containing a final **return** statement and declaring an integer variable for increment by a loop

```
int main()
{
  int number ;
  // Add more statements here.
  return 0 ;
}
```

3 In the main function, insert **try** and **catch** blocks to handle a "bad number" exception

```
try
{        }
catch ( int num )
{        }
```

4 In the **try** block, insert a loop to increment the variable

```
for ( number = 1 ; number < 21 ; number++ )
{
  if (number > 4 ) throw ( number ) ;
  else
  cout << "Number: " << number << endl ;
}
```

5 In the **catch** block, insert an exception handler statement

```
cout << "Exception at: " << num << endl ;
```

6 Save, compile, and run the program to see the thrown exception get caught by the **catch** block

Don't forget

When an exception occurs, control passes to the **catch** block – in this example, the loop does not complete.

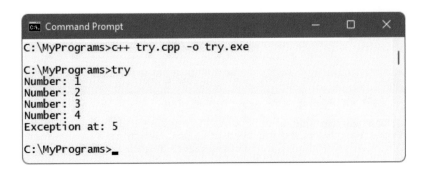

```
C:\MyPrograms>c++ try.cpp -o try.exe

C:\MyPrograms>try
Number: 1
Number: 2
Number: 3
Number: 4
Exception at: 5

C:\MyPrograms>
```

Recognizing exceptions

When a program exception occurs within a **try** block, an "exception" object is automatically thrown. A reference to the exception can be passed to the associated **catch** block in the parentheses after the **catch** keyword. This specifies the argument to be an **exception** type, and a chosen exception name prefixed by the **&** reference operator – for example, **exception &error**.

Once an exception reference is passed to a **catch** block, an error description can be retrieved by the exception's **what()** function:

what.cpp

1 Start a new program by specifying the C++ library classes to include, and a namespace prefix to use
```
#include <string>                    // Include string support.
#include <iostream>
using namespace std ;
```

2 Add a main function containing a final **return** statement and declaring an initialized **string** variable
```
int main()
{
  string lang = "C++" ;
  // Add more statements here.
  return 0 ;
}
```

3 In the main function, insert a **try** block containing a statement attempting to erase part of the **string** variable
```
try { lang.erase( 4, 6 ) ; }
```

4 Next, insert a **catch** block containing a statement to send a description to standard error output by the **cerr** function
```
catch ( exception &error )
{ cerr << "Exception: " << error.what() << endl ; }
```

5 Save, compile, and run the program to see the error description of the caught exception

Don't forget

The error description will vary for different compilers – the Visual C++ compiler describes the exception error in this example as an "invalid string position".

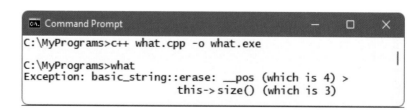

```
Command Prompt                              —    □    ×
C:\MyPrograms>c++ what.cpp -o what.exe

C:\MyPrograms>what
Exception: basic_string::erase: __pos (which is 4) >
                            this->size() (which is 3)
```

The C++ **<stdexcept>** library defines a number of exception classes. Its base class is named "exception", from which other classes are derived, categorizing common exceptions.

Each of the exception classes are listed in the table below, illustrating their relationship and describing their exception type:

Exception class:	Description:
exception	General exception
— bad_alloc	– failure allocating storage
— bad_cast	– failure casting data type
— bad_typeid	– failure referencing typeid
— logic_error	Logic exception
domain_error	– invalid domain
invalid_argument	– invalid argument in call
length_error	– invalid length for container
out_of_range	– invalid element range
— runtime_error	Runtime exception
range_error	– invalid range request
overflow_error	– invalid arithmetic request

When the **<stdexcept>** library is made available to a program, by adding an **#include <stdexcept>** preprocessor directive, the exception classes can be used to identify the type of exception thrown to a **catch** block.

The specific exception class name can appear, in place of the general **exception** type, within the **catch** block's parentheses.

Multiple **catch** blocks can be used in succession, much like **case** statements in a **switch** block, to handle several types of exception.

Additionally, exceptions can be produced manually by the **throw** keyword. This can be used to create any of the **logic_error** and **runtime_error** exceptions in the table above. Optionally, a custom error message can be specified for manual exceptions, which can be retrieved by its **what()** function.

Hot tip

The **cout** function sends data to standard output, whereas the **cerr** function sends error data to standard error output. These are simply two different data streams.

Hot tip

The example described on page 94 demonstrates the use of standard exceptions, and you will find more on references in Chapter 6.

Handling errors

Exception type information can be provided by including the C++ standard **<typeinfo>** library. Its **typeid()** function accepts an exception argument so its **name()** function can return the type name:

except.cpp

1 Start a new program by specifying the C++ library classes to include, and a namespace prefix to use
```
#include <stdexcept>   // Support standard exceptions.
#include <typeinfo>    // Support type information.
#include <fstream>
#include <string>
#include <iostream>
using namespace std ;
```

2 Add a main function containing a final **return** statement, two initialized variable declarations, and a statement outputting a text message
```
int main()
{
  string lang = "C++" ;
  int num = 1000000000 ; // One billion.
  // Try-catch block goes here.
  cout << "Program continues..." << endl ;
  return 0 ;
}
```

3 In the main function, insert a **try** block containing a statement attempting to replace part of the **string** value
```
try { lang.replace( 100, 1 , "C" ) ; }
```

The **out_of_range** error occurs because the **replace()** function is trying to begin erasing at the 100th character, but the **string** variable has only three characters.

4 After the **try** block, add a **catch** block to handle a range exception then save, compile, and run the program
```
catch ( out_of_range &e )
{
  cerr << "Range Exception: " << e.what() << endl ;
  cerr << "Exception Type: " << typeid( e ).name() ;
  cerr << endl << "Program terminated." << endl ;
  return -1 ;
}
```

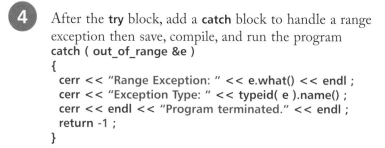

```
Command Prompt                                          —    □    ×

C:\MyPrograms>c++ except.cpp -o except.exe

C:\MyPrograms>except
Range Exception: basic_string::replace: __pos (which is 100
                              this->size() (which is 3)
Exception Type: St12out_of_range
Program terminated.
```

5 Replace the statement in the **try** block with one attempting to resize the **string** variable
lang.resize(3 * num) ;

6 After the **catch** block, add a second **catch** block to handle general exceptions
```
catch ( exception &e )
{
  cerr << "Exception: " << e.what() << endl ;
  cerr << "Exception Type: " << typeid( e ).name() << endl ;
}
```

7 Save, compile, and run the program again to see the exception handled by the second **catch** block

```
Command Prompt                          —    □    X

C:\MyPrograms>except
Exception: basic_string::_M_replace_aux
Exception Type: St12length_error
Program continues...

C:\MyPrograms>_
```

Beware

The order of **catch** blocks can be important – placing the **exception** error handler before the **out_of_range** error handler would allow an **out_of_range** error to be handled by the (higher level) **exception** handler.

8 Replace the statement in the **try** block with one attempting to open a non-existent file
```
ifstream reader( "nonsuch.txt" ) ;
if ( ! reader ) throw logic_error( "File not found" ) ;
```

9 Save, compile, and run the program once more to see the exception handled again by the second **catch** block, and the specified custom error message

```
Command Prompt                          —    □    X

C:\MyPrograms>c++ except.cpp -o except.exe

C:\MyPrograms>except
Exception: File not found
Exception Type: St11logic_error
Program continues...

C:\MyPrograms>_
```

Hot tip

An exception object is typically given the name "e" – as seen here.

95

Summary

- The C++ **\<fstream\>** library provides functions for working with files as **ifstream** input or **ofstream** output stream objects.

- Upon completion, a stream's **close()** function should be called.

- File modes can be used to control the behavior of a stream.

- An input stream's **get()** function reads one character at a time.

- The **getline()** function can be used to read a line at a time from an input stream.

- Optionally, the **getline()** function can have a third argument specifying a delimiter character at which to stop reading.

- Insertion operators can be used with the **cin** and **cout** functions to modify their behavior.

- The **cout.width()** function sets the width of the output stream.

- The **cout.fill()** function specifies a character to occupy any empty portion of the output.

- The **cout.precision()** function determines how many decimal places to display when the output is a floating point number.

- A badly performing program may contain syntax errors, logic errors, or exception errors.

- A **try** block can be used to enclose statements that, under certain conditions, may cause an exception error.

- A **catch** block can be used to handle exception errors produced in its associated **try** block.

- Exception errors that occur in a **try** block are automatically thrown to the associated **catch** block, or can be manually thrown using the **throw()** function.

- The C++ **\<stdexcept\>** library defines a number of exception classes that categorize common exceptions, and the **\<typeinfo\>** library provides exception type information.

6 Pointing to data

This chapter demonstrates how to produce efficient C++ programs, utilizing pointers and references.

0x01   0x02   0x03

Understanding data storage

In order to understand C++ pointers it is helpful to understand how data is stored on your computer. Envision the computer's memory as a very long row of sequentially-numbered slots, which can each contain one byte of data. When a variable is declared in a program, the machine reserves a number of slots at which to store data assigned to that variable. The number of slots it reserves depends upon the variable's data type. When the program uses the variable name to retrieve stored data, it actually addresses the number of the variable's first reserved slot.

Comparison can be drawn to a long row of sequentially-numbered houses that can each accommodate a different family. Any family can be explicitly referenced by addressing their house number.

The slot (house) numbers in computer memory are expressed in hexadecimal format and can be revealed by the **&** reference operator.

address.cpp

1 Start a new program by specifying the C++ library classes to include, and a namespace prefix to use
```
#include <string>
#include <iostream>
using namespace std ;
```

2 Add a main function containing a final **return** statement and declaring three initialized variables
```
int main()
{
  int num = 100 ;
  double sum = 0.0123456789 ;
  string text = "C++ Fun" ;
  // Add more statements here.
  return 0 ;
}
```

3 In the main function, insert statements to output the memory address of the first slot of each variable
```
cout << "Integer variable starts at: " << &num << endl ;
cout << "Double variable starts at: " << &sum << endl ;
cout << "String variable starts at: " << &text << endl ;
```

4 Save, compile, and run the program to see the three memory addresses

```
C:\MyPrograms>c++ address.cpp -o address.exe

C:\MyPrograms>address
Integer variable starts at: 0x61ff08
Double variable starts at: 0x61ff00
String variable starts at: 0x61fee8
```

Once memory space has been reserved by a variable declaration, a value of the appropriate data type can be stored there using the = assignment operator. For example, **num = 100** takes the value on its right (**100**) and puts it in the memory referenced by the named variable on its left (**num**).

The operand to the left of the = assignment operator is called its "L-value" and the operand to its right is called its "R-value". Consider the "L" in L-value to mean "Location" and consider the "R" in R-value to mean "Read".

One important rule in C++ programming insists that an R-value cannot appear to the left of the = assignment operator, but an L-value may appear on either side. Code that places an R-value to the left of an = assignment operator will not compile:

5 Just before the **return** statement, insert statements placing R-values incorrectly to the left of assignment operators
200 = num ;
5.5 = sum ;
"Bad assignments" = text ;

6 Save, and attempt to recompile the program to see the errors caused by incorrectly placed R-values

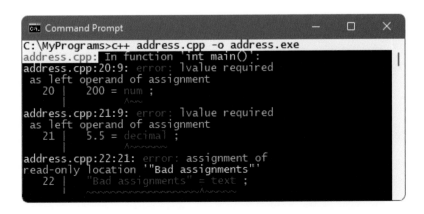

The location addresses are dynamically allocated – so will vary from those in this screenshot.

99

L-values are containers but R-values are data.

Getting values with pointers

Pointers are a useful part of efficient C++ programming – they are simply variables that store the memory address of other variables.

Pointer variables are declared in the same way that other variables are declared, but the data type is suffixed by a "*" character. This denotes, in a declaration, that the variable will be a pointer. Always remember that the pointer's data type must match that of the variable to which it points.

A pointer variable is initialized by assigning it the memory address of another variable, using the & reference operator. The assignment can be made either in the declaration statement, or in a separate statement after the declaration. Referencing a pointer variable by its name alone will simply reveal the memory address that it contains.

After a pointer variable has been initialized, either in the declaration or by a subsequent assignment, it "points" to the variable at the address that it contains. Usefully, this means that the value of the variable to which it points can be referenced by prefixing the pointer name with the * dereference operator:

deref.cpp

1 Start a new program by specifying the C++ library classes to include, and a namespace prefix to use
```
#include <iostream>
using namespace std ;
```

2 Add a main function containing a final **return** statement and declaring two regular initialized integer variables
```
int main()
{
  int a = 8 , b = 16 ;
  // Add more statements here.
  return 0 ;
}
```

3 In the main function, insert a statement to declare, and initialize a pointer with the memory address of the first integer variable
```
int* aPtr = &a ;          // aPtr assigned address of a.
```

4 Insert a statement to declare a second pointer, then initialize it with the address of the second integer variable

```
int* bPtr ;            // bPtr declared.
bPtr = &b ;            // bPtr assigned address of b.
```

The * dereference operator is alternatively known as the "indirection" operator.

5 Next, insert statements to output the actual memory address of each pointer

```
cout << "Addresses of pointers..." << endl ;
cout << "aPtr: " << &aPtr << endl ;
cout << "bPtr: " << &bPtr << endl << endl ;
```

6 Now, insert statements to output the memory address stored inside each pointer

```
cout << "Values in pointers..." << endl ;
cout << "aPtr: " << aPtr << endl ;
cout << "bPtr: " << bPtr << endl << endl ;
```

7 Finally, insert statements to output the values stored at the memory address stored in each pointer – the value of the variables to which the pointers point

```
cout << "Values in addresses pointed to..." << endl ;
cout << "a: " << *aPtr << endl ;
cout << "b: " << *bPtr << endl ;
```

8 Save, compile, and run the program to see the pointer addresses and the stored values

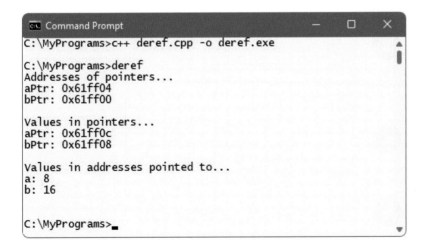

```
C:\MyPrograms>c++ deref.cpp -o deref.exe

C:\MyPrograms>deref
Addresses of pointers...
aPtr: 0x61ff04
bPtr: 0x61ff00

Values in pointers...
aPtr: 0x61ff0c
bPtr: 0x61ff08

Values in addresses pointed to...
a: 8
b: 16

C:\MyPrograms>
```

The memory addresses are dynamically allocated – so will vary from those in this screenshot.

Doing pointer arithmetic

Once a pointer variable has been initialized with a memory address, it can be assigned another address or changed using pointer arithmetic.

The **++** increment operator or the **--** decrement operator will move the pointer along to the next or previous address for that data type – the larger the data type, the bigger the jump.

Even larger jumps can be achieved using the **+=** and **-=** operators.

Pointer arithmetic is especially useful with arrays, because the elements in an array occupy consecutive memory addresses. Assigning just the name of an array to a pointer automatically assigns it the address of the first element. Incrementing the pointer by 1 moves the pointer along to the next element.

point.cpp

1 Start a new program by specifying the C++ library classes to include, and a namespace prefix to use
#include <iostream>
using namespace std ;

2 Add a main function containing a final **return** statement and declaring an initialized integer array of 10 elements
int main()
{
  int nums[] = { 1, 2, 3, 4, 5, 6, 7, 8, 9, 10 } ;
  // Add more statements here.
  return 0 ;
}

3 In the main function, insert a statement to declare a pointer, initialized with the memory address of the first element in the integer array
int* ptr = nums ;

4 Next, insert a statement to output the memory address of the first element of the integer array, and its value
cout << endl << "ptr at: " << ptr << " gets: "<< *ptr ;

5 Now, increment the pointer and output its new memory address – that of the second element in the integer array
ptr++ ;
cout << endl << "ptr at: " << ptr << " gets: "<< *ptr ;

...cont'd

6 Increment the pointer again, and output its new memory
address – that of the third element in the integer array

```
ptr++ ;
cout << endl << "ptr at: " << ptr << " gets: " << *ptr ;
```

The *=, /=, and %=
operators cannot be
used to move a pointer.

7 Decrement the pointer by two places and output its
memory address – that of the first element in the array

```
ptr -= 2 ;
cout << endl << "ptr at: " << ptr << " gets: " << *ptr ;
cout << endl ;
```

8 Now, insert a loop to output the value stored in each
element of the integer array

```
for ( int i = 0 ; i < 10 ; i++ )
{
        cout << endl << "Element: " << i ;
        cout << "    Value: " << *ptr ;
        ptr++ ;
}
cout << endl ;
```

9 Save, compile, and run the program to see the pointer
addresses and the stored values

```
Command Prompt                              —    □    ×

C:\MyPrograms>c++ point.cpp -o point.exe

C:\MyPrograms>point

ptr at: 0x61fee0 gets: 1
ptr at: 0x61fee4 gets: 2
ptr at: 0x61fee8 gets: 3
ptr at: 0x61fee0 gets: 1

Element: 0      Value: 1
Element: 1      Value: 2
Element: 2      Value: 3
Element: 3      Value: 4
Element: 4      Value: 5
Element: 5      Value: 6
Element: 6      Value: 7
Element: 7      Value: 8
Element: 8      Value: 9
Element: 9      Value: 10

C:\MyPrograms>_
```

The name of an array
acts like a pointer to its
first element.

Passing pointers to functions

Pointers can access the data stored in the variable to which they point using the * dereference operator. This can also be used to change the stored data by assigning a new value of the appropriate data type.

Additionally, pointers can be passed to functions as arguments – with a subtly different effect to passing variables as arguments:

● When variables are passed to functions, their data is passed "by value" to a local variable inside the function – so that the function operates on a <u>copy</u> of the original value.

● When pointers are passed to functions, their data is passed "by reference" – so that the function operates on the <u>original</u> value.

The benefit of passing by reference allows functions to directly manipulate variable values declared within the calling function.

fnptr.cpp

1 Start a new program by specifying the C++ library classes to include, and a namespace prefix to use
```
#include <iostream>
using namespace std ;
```

2 After the preprocessor instructions, declare two function prototypes to each accept a single pointer argument
```
void writeOutput ( int* ) ;
void computeTriple ( int* ) ;
```

3 Add a main function containing a final **return** statement and declaring an initialized regular integer variable
```
int main()
{
  int num = 5 ;
  // Add more statements here.
  return 0 ;
}
```

4 In the main function, insert a second variable declaration that initializes a pointer with the address of the regular integer variable
```
int* ptr = &num ;
```

5 After the main function block, define a function to output the current value of a variable to which a pointer points

```
void writeOutput( int* value )
{
  cout << "Current value: " << *value << endl ;
}
```

The function prototype and definition must both contain a pointer argument.

6 Define another function to multiply the current value of a variable to which a pointer points

```
void computeTriple( int* value )
{
  *value *= 3 ;  // Multiply and assign dereferenced value.
}
```

7 In the main function, pass a pointer argument to a function to output the variable value to which it points

```
writeOutput( ptr ) ;
```

8 Next, use the pointer to increase the variable value, then display its new value

```
*ptr += 15 ;   // Add and assign a dereferenced value.
writeOutput( ptr ) ;
```

9 Now, pass a pointer argument to a function to multiply the variable to which it points, then display its new value

```
computeTriple( ptr ) ;
writeOutput( ptr ) ;
```

Hot tip

Functions that operate directly on variables within the calling function need no **return** statement.

10 Save, compile, and run the program to see the computed values output

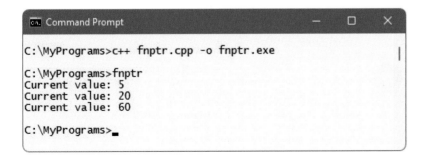

```
C:\MyPrograms>c++ fnptr.cpp -o fnptr.exe

C:\MyPrograms>fnptr
Current value: 5
Current value: 20
Current value: 60

C:\MyPrograms>
```

105

Making arrays of pointers

A variable of the regular **char** data type can be assigned a single character value, but a pointer to a constant **char** array can usefully be assigned a string of characters. The string is actually stored as an array of characters, with one character per element, but referencing the **char** pointer will automatically retrieve the entire string.

This ability to retrieve a string value from a **char** pointer using just its variable name resembles the way that a string can be retrieved from a regular **char** array using its variable name.

Multiple strings can be stored in a constant **char** pointer array, with one string per element. You can even store multiple **char** pointer arrays in a "master" **char** pointer array – one array per element.

arrptr.cpp

1 Start a new program by specifying the C++ library classes to include, and a namespace prefix to use
```
#include <iostream>
using namespace std ;
```

2 Add a main function containing a final **return** statement and declaring two initialized variables – a regular character array, and a character pointer with identical **string** values
```
int main()
{
  char letters[ 8 ] = { 'C', '+' , '+' , ' ' , 'F', 'u', 'n', '\0' } ;
  const char* text = "C++ Fun" ;
  // Add more statements here.
  return 0 ;
}
```

3 In the main function, insert statements to declare, and initialize two further character pointer variables, with unique **string** values
```
const char* term = "Element:" ;
const char* lang = "C++" ;
```

4 Next, insert a statement to declare a character pointer array, initialized with three **string** values
```
const char* ap1[ 3 ] = { "Great " , "Program" , "Code   " } ;
```

Don't forget

Character values must be enclosed in single quotes, but **string** values must be enclosed in double quotes – even when they are being assigned to a **char** pointer.

5 Now, insert a statement to declare a second character pointer array, initialized with three string values – making one of the pointer variables its first element value
`const char* ap2[ 3 ] = { lang , "is " , "Fun" } ;`

To include a space in a **char** array, the assignment must have a space between the quotes as ' ' – two single quotes together (") is regarded as an empty element and causes a compiler error.

6 Declare two "master" character pointer arrays, each initialized with three elements of the **char** pointer arrays
`const char* ap3[ 3 ] = { ap2[ 0 ] , ap2[ 1 ] , ap1[ 0 ] } ;`
`const char* ap4[ 3 ] = { ap1[ 2 ] , ap2[ 1 ] , ap2[ 2 ] } ;`

7 After the declarations, insert statements to output the identical **string** values of the first two variables
`cout << letters << endl ;`
`cout << text << endl ;`

8 Next, insert a loop containing a statement to output the value within a character pointer and the iteration number
```
for ( int i = 0 ; i < 3 ; i++ )
{
  cout << term << i << "   " ;
}
```

9 Within the loop block, insert statements to output each element value of the four character pointer arrays
```
cout << ap1[ i ] << "   " ;
cout << ap2[ i ] << "   " ;
cout << ap3[ i ] << "   " ;
cout << ap4[ i ] << endl ;
```

Remember that the final element of a **char** array must contain the special \0 character to designate that array as a string.

10 Save, compile, and run the program to see the character string output

```
Command Prompt                               —  □  ×

C:\MyPrograms>c++ arrptr.cpp -o arrptr.exe

C:\MyPrograms>arrptr
C++ Fun
C++ Fun
Element:0    Great     C++    C++    Code
Element:1    Program   is     is     is
Element:2    Code      Fun    Great    Fun

C:\MyPrograms>_
```

Referencing data

A C++ "reference" is an alias for a variable or an object in a program. A reference <u>must</u> be initialized within its declaration, by assigning it the name of the variable or object to which it refers. From then on, the reference acts as an alternative name for the item to which it refers – anything that happens to the reference is really implemented on the variable or object to which it refers.

A reference declaration first states its data type, matching that of the item to which it will refer, suffixed by an **&** character denoting that variable will be a reference, and a chosen name. Finally, the declaration uses the **=** operator to associate a variable or object.

Traditionally, a reference is named with the name of the associated variable or object, but with an uppercase first letter and the whole name prefixed by an "r". For example, a declaration to create a reference to an integer variable named "num" looks like this:

int& rNum = num ;

Note that the purpose of the **&** reference operator is context-sensitive so that it declares a reference when used as an L-value, on the left side of the **=** operator, otherwise it returns a memory address when used as an R-value.

A reference is such a true alias to its associated item that querying its memory address returns the address of its associated item – there is no way to discover the address of the reference itself.

Beware

Once a reference has been created, it will always refer to the item to which it was initialized – a different item cannot be assigned to that reference.

108

C++

ref.cpp

1 Start a new program by specifying the C++ library classes to include, and a namespace prefix to use
```
#include <iostream>
using namespace std ;
```

2 Add a main function containing a final **return** statement and declaring two variables – a regular integer variable and a reference to that variable
```
int main()
{
  int num ;
  int& rNum = num ;
  // Add more statements here.
  return 0 ;
}
```

...cont'd

3 In the main function, insert a statement assigning an initial value to the integer variable via its reference
rNum = 400 ;

4 Next, insert statements to output the stored value, both directly and via its reference
cout << "Value direct: " << num << endl ;
cout << "Value via reference: " << rNum << endl ;

5 Now, insert statements to output the memory address, both directly and via its reference
cout << "Address direct: " << &num << endl ;
cout << "Address via reference: " << &rNum << endl ;

6 Insert a statement to manipulate the value stored in the variable via its reference
rNum *= 2 ;

7 Once more, output the stored value, both directly and via its reference
cout << "Value direct: " << num << endl ;
cout << "Value via reference: " << rNum << endl ;

8 Save, compile, and run the program to see the values and memory address

The compiler decides how to use the **&** reference operator according to its context.

A reference is always an alias for the item associated in its declaration statement.

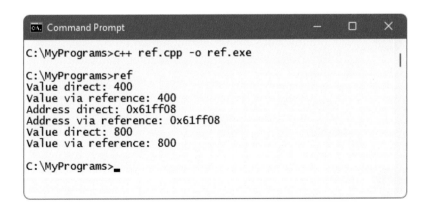

```
Command Prompt                                    —    □    ×

C:\MyPrograms>c++ ref.cpp -o ref.exe                         |

C:\MyPrograms>ref
Value direct: 400
Value via reference: 400
Address direct: 0x61ff08
Address via reference: 0x61ff08
Value direct: 800
Value via reference: 800

C:\MyPrograms>_
```

Passing references to functions

References provide access to the data stored in the variable to which they refer, just like the variable itself, and can be used to change the stored data by assigning a new value of the appropriate data type.

Additionally, references can, like pointers, be passed to functions as arguments:

- When variables are passed to functions, their data is passed "by value" to a local variable inside the function – so that the function operates on a <u>copy</u> of the original value.

- When references are passed to functions, their data is passed "by reference" – so the function operates on the <u>original</u> value.

The benefit of passing by reference allows functions to directly manipulate variable values declared within the calling function.

fnref.cpp

1 Start a new program by specifying the C++ library classes to include, and a namespace prefix to use
```
#include <iostream>
using namespace std ;
```

2 After the preprocessor instructions, declare two function prototypes to each accept a single reference argument
```
void writeOutput ( int& ) ;
void computeTriple ( int& ) ;
```

3 Add a main function containing a final **return** statement and declaring an initialized regular integer variable
```
int main()
{
  int num = 5 ;
  // Add more statements here.
  return 0 ;
}
```

Don't forget

This example may seem familiar, as it recreates the example on page 104 – but replaces pointers with references.

4 In the main function, insert a second variable declaration, initializing a reference as an alias of the integer variable
```
int& ref = num ;
```

5 After the main function block, define a function to output the current value of a variable to which a reference refers

```
void writeOutput( int& value )
{
  cout << "Current value: " << value << endl ;
}
```

The function prototype and definition must both contain a reference argument.

6 Define another function to multiply the current value of a variable to which a reference refers

```
void computeTriple( int& value )
{
  value *= 3 ;   // Multiply and assign a referenced value.
}
```

7 In the main function, pass a reference argument to a function to output the variable value to which it refers

```
writeOutput( ref ) ;
```

8 Next, use the reference to increase the variable value, then display its new value

```
ref += 15 ;      // Add and assign a referenced value.
writeOutput( ref ) ;
```

Functions that operate directly on variables within the calling function need no **return** statement.

9 Now, pass a reference argument to a function to multiply the variable to which it refers, then display its new value

```
computeTriple( ref ) ;
writeOutput( ref ) ;
```

10 Save, compile, and run the program to see the computed values output

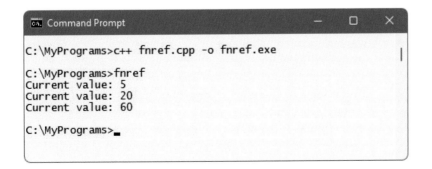

Comparing pointers & references

Pointers and references can both be used to refer to variable values and to pass them to functions by reference rather than by value. Technically, passing by reference is more efficient than passing by value, so the use of pointers and references is to be encouraged.

The decision whether to use a pointer or a reference is determined by the program requirements. C++ programmers generally prefer to use references wherever possible, as they are easier to use and easier to understand than pointers. References must, however, obey certain rules that can make the use of pointers necessary:

Rule:	References:	Pointers:
Can be declared without initialization	No	Yes
Can be reassigned	No	Yes
Can contain a 0 (null) value	No	Yes
Easiest to use	Yes	No

As a general rule, the choice between using a reference or a pointer can be determined by following these guidelines:

● If you don't want to initialize in the declaration, use a pointer;

or

● If you want to be able to reassign another variable, use a pointer;

otherwise

● Always use a reference.

Pointers are more flexible than references, however, and can even point to functions. In this case, the pointer declaration must precisely match the return data type and arguments to those of the function to which it points. Additionally, the pointer name must be enclosed within parentheses in the declaration to avoid compiler errors. The function pointer can then be assigned a function by name, so that function can be called via the pointer.

Don't forget

A reference must be initialized in the declaration to refer to a variable or object – then always refers to that item.

1 Start a new program by specifying the C++ library classes to include, and a namespace prefix to use

```
#include <iostream>
using namespace std ;
```

pref.cpp

2 After the preprocessor instructions, define an inline function to output the total of two passed arguments

```
inline void add ( int& a, int* b )
{ cout << "Total: " << ( a + *b ) << endl ; }
```

3 Add a main function containing a final **return** statement and declarations creating two regular integer variables, one reference, and two pointers

```
int main()
{
  int num = 100 , sum = 500 ;
  int& rNum = num ;
  int* ptr = &num ;
  void (* fn ) ( int& a, int* b ) = add ;
  // Add more statements here.
  return 0 ;
}
```

4 In the main function, insert statements to output the first integer variable values via the reference and pointer

```
cout << "Reference: " << rNum << endl ;
cout << "Pointer: " << *ptr << endl ;
```

5 Now, assign the second integer variable to the pointer and output its value via the pointer, then call the function pointer to output the sum of the variable values

```
ptr = &sum ;
cout << "Pointer now: " << *ptr << endl ;
fn( rNum , ptr ) ;
```

6 Save, compile, and run the program to see the output

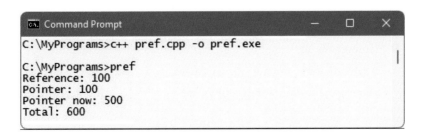

```
Command Prompt                              —  □  ×

C:\MyPrograms>c++ pref.cpp -o pref.exe

C:\MyPrograms>pref
Reference: 100
Pointer: 100
Pointer now: 500
Total: 600
```

113

Hot tip

C programmers tend to put the **&** and ***** characters before the variable names, but in C++ it is usual to put them after the data type – as the feature is a property of the data type, not the name.

Summary

- Data is stored in computer memory within sequentially numbered addresses.

- Operands to the left of an = operator are L-values, and those to its right are R-values.

- An R-value may only appear to the right of an = operator.

- A pointer is a variable that stores the memory address of another variable – that to which it points.

- The * character appears as an L-value in a pointer declaration, indicating that the statement will create a pointer variable.

- Once declared, the * dereference operator can be used to reference the value within a variable to which a pointer points.

- Pointer arithmetic can be used to iterate through the values stored in array elements.

- A variable is passed to a function by value, whereas pointers and references are passed by reference.

- Passing by reference allows the receiving function to directly manipulate variables declared within the calling function.

- A string value can be assigned to a pointer of the **char** data type, and the whole string retrieved using the pointer name.

- Each element in a pointer array can store data or a pointer.

- A reference is not a variable, but merely an alias for a variable.

- The **&** character appears as an L-value in a reference declaration, indicating that the statement will create an alias.

- The **&** reference operator can be used to reference the memory address stored within a pointer.

- References are easier to use than pointers but, unlike pointers, a reference must always be initialized in its declaration and can never be assigned a different variable.

7 Creating classes and objects

This chapter introduces the topics of encapsulation and inheritance – the first two principles of C++ Object Oriented Programming.

Encapsulating data

A class is a data structure that can contain both variables and functions in a single entity. These are collectively known as its "members", and the functions are known as its "methods".

Access to class members from outside the class is controlled by "access specifiers" in the class declaration. Typically, these will deny access to the variable members, but allow access to methods that can store and retrieve data from those variable members. This technique of "data hiding" ensures that stored data is safely encapsulated within the class variable members, and is the first principle of Object Oriented Programming (OOP).

A class declaration begins with the **class** keyword, followed by a space, then a programmer-specified name – adhering to the usual C++ naming conventions, but beginning in uppercase. Next, come the access specifiers and class members, contained within a pair of braces. Every class declaration must end with a semicolon after the closing brace – so the class declaration syntax looks like this:

```
class ClassName
{
  access specifier :
    member1 ;
    member2 ;
  access specifier :
    member3 ;
    member4 ;
} ;
```

An access specifier may be any one of the keywords **public**, **private**, or **protected** to specify access rights for its listed members:

● Public members are accessible from any place where the class is visible.

● Private members are accessible only to other members of the same class.

● Protected members are accessible only to other members of the same class and to members of classes derived from that class.

By default, all class members have private access – so any members that appear in the class declaration without an access specifier will have private access.

Derived classes, which use the **protected** access specifier, are introduced later in this chapter.

Any real-world object can be defined by its attributes and by its actions. For example, a dog has attributes such as age, weight, and color, and actions it can perform such as bark. The class mechanism in C++ provides a way to create a virtual dog object within a program, where the variable members of a class can represent its attributes, and the methods represent its actions:

```
class Dog
{
  private: // The default access level.
    int age, weight ;
    string color ;
  public :
    void bark() ;
    // ... Plus methods to store/retrieve data.
} ;
```

It is important to recognize that a class declaration only defines a data structure – in order to create an object you must declare an "instance" of that data structure. This is achieved in just the same way that instances are declared of regular C++ data types:

```
int num ;        // Creates an instance named "num".
                 // of the regular C++ int data type.

Dog fido ;       // Creates an instance named "fido".
                 // of the programmer-defined Dog data structure.
```

Alternatively, an instance object can be created by specifying its name between the class declaration's closing brace and its final semicolon. Multiple instances can be created this way, by specifying a comma-separated list of object names. For example, the class declaration listed below creates four instance objects of the Dog class named "fido", "pooch", "rex", and "sammy".

```
class Dog
{
    int age, weight ;
    string color ;
  public:
    void bark() ;
    // ... Plus methods to store/retrieve data.
} fido, pooch, rex, sammy ;
```

The principle of encapsulation in C++ programming describes the grouping together of data and functionality in class members – age, weight, color attributes and bark action in the Dog class.

While a program class cannot perfectly emulate a real-word object, the aim is to encapsulate all relevant attributes and actions.

It is conventional to begin class names with an uppercase character and object names with lowercase.

117

Creating an object

In order to assign and retrieve data from private members of a class, special public accessor methods must be added to the class. These are "setter" methods, to assign data, and "getter" methods, to retrieve data. Accessor methods are often named as the variable they address, with the first letter made uppercase, and prefixed by "set" or "get" respectively. For example, accessor methods to address an **age** variable may be named **setAge()** and **getAge()**.

object.cpp

1 Start a new program by specifying the C++ library classes to include, and a namespace prefix to use
```
#include <string>
#include <iostream>
using namespace std ;
```

Don't forget

Members declared before an access specifier are **private** by default, and remember to add a final semicolon after each class declaration.

2 Declare a class named "Dog"
```
class Dog
{

} ;
```

3 Between the braces of the Dog class declaration, declare three **private** variable members
```
int age, weight ;
string color ;
```

4 After the **private** variables, add a **public** access specifier
```
public:
```

5 Begin the **public** members list by adding a method to output a **string** when called
```
void bark() { cout << "WOOF!" << endl ; }
```

Hot tip

In the class declaration, notice that all methods are declared **public** and all variables are declared **private**. This notion of "public interface, private data" is a key concept when creating classes.

6 Add **public** setter methods – to assign individual values to each of the **private** variables
```
void setAge ( int yrs ) { age = yrs ; }
void setWeight ( int lbs ) { weight = lbs ; }
void setColor ( string hue ) { color = hue ; }
```

7 Add **public** getter methods – to retrieve individual values from each of the **private** variables
```
int getAge()    { return age ; }
int getWeight() { return weight ; }
string getColor() { return color ; }
```

8 After the Dog class declaration, declare a main method containing a final **return** statement

```
int main()
{
    // Program code goes here.
    return 0 ;
}
```

9 Between the braces of the main method, declare an instance of the Dog class named "fido"

```
Dog fido ;
```

Fido

10 Add statements calling each setter method to assign data

```
fido.setAge( 3 ) ;
fido.setWeight( 15 ) ;
fido.setColor( "brown" ) ;
```

11 Add statements calling each getter method to retrieve the assigned values

```
cout << "Fido is a " << fido.getColor() <<
                            " dog" << endl ;
cout << "Fido is " << fido.getAge() <<
                            " years old" << endl ;
cout << "Fido weighs " << fido.getWeight() <<
                            " pounds" << endl ;
```

Hot tip

12 Now, add a call to the regular output method

```
fido.bark() ;
```

This program will get modified over the next few pages as new features are incorporated.

13 Save, compile, and run the program to see the output

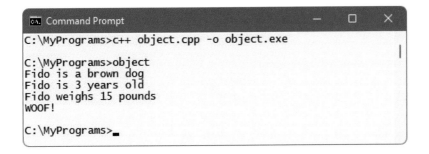

```
C:\MyPrograms>c++ object.cpp -o object.exe

C:\MyPrograms>object
Fido is a brown dog
Fido is 3 years old
Fido weighs 15 pounds
WOOF!

C:\MyPrograms>
```

Creating multiple objects

A program can easily create multiple objects, simply by declaring multiple instances of a class, and each object can have unique attributes by assigning individual values with its setter methods.

It is often convenient to combine the setter methods into a single method that accepts arguments for each private variable. This means that all values can be assigned with a single statement in the program, but the method will contain multiple statements.

The class declaration in the previous example contains short methods of just one line, which are created "inline" – entirely within the class declaration block. Where methods have more than two lines, they should not be created inline, but should instead be declared as a "prototype" in the class declaration block and defined separately – after the class declaration. The definition must prefix the method name with the class name and the scope resolution operator :: to identify the class containing its prototype.

multiple.cpp

1 Rename a copy of the previous example "object.cpp" as a new program "multiple.cpp"

2 In the Dog class declaration, replace the three setter methods with a single combined setter prototype that specifies the argument data types – but not their names
void setValues (int, int, string) ;

3 After the Dog class declaration, add a definition block for the prototype using the :: scope resolution operator to identify the class in which it resides
void Dog::setValues (int age, int weight, string color)
{

}

Don't
forget

Note that a prototype is a statement – so it must end with a semicolon.

Notice that, for easy identification, the arguments are named with the same names as the variables to which they will be assigned. Where a class method definition has an argument of the same name as a class member, the **this ->** class pointer can be used to explicitly refer to the class member. For example, **this -> age** refers to the class member variable, whereas **age** refers to the argument.

4 In the method definition block, insert three statements to assign values from passed arguments to class variables
```
this -> age = age ;
this -> weight = weight ;
this -> color = color ;
```

5 Between the braces of the main method, replace the calls to the three setter methods by a single call to the combined setter method – passing three arguments
```
fido.setValues( 3, 15, "brown" ) ;
```

6 In the main method, declare a second instance of the Dog class named "pooch"
```
Dog pooch ;
```

7 Add a second call to the combined setter method – passing three arguments for the new object
```
pooch.setValues( 4, 18, "gray" ) ;
```

8 Add statements calling each getter method to retrieve the assigned values
```
cout << "Pooch is a " << pooch.getAge() ;
cout << " year old " << pooch.getColor() ;
cout << " dog who weighs " << pooch.getWeight() ;
cout << " pounds ." ;
```

9 Now, add second call to the regular output method
```
pooch.bark() ;
```

10 Save, compile, and run the program to see the output

Where the argument name and class member names are different, the **this ->** class pointer is not needed in the setter method definitions.

121

```
Command Prompt                    —    □    ✕

C:\MyPrograms>c++ multiple.cpp -o multiple.exe

C:\MyPrograms>multiple
Fido is a brown dog
Fido is 3 years old
Fido weighs 15 pounds
WOOF!
Pooch is a 4 year old gray dog who weighs 18 pounds.WOOF!

C:\MyPrograms>_
```

Fido

Pooch

Initializing class members

Class variable members can be initialized by a special "constructor" method that is called whenever an instance of the class is created. The constructor method is always named exactly as the class name, and requires arguments to set the initial value of class variables.

When a constructor method is declared, an associated "destructor" method should also be declared – that is called whenever an instance of the class is destroyed. The destructor method is always named as the class name, prefixed by a ~ tilde character.

Constructor and destructor methods have no return value, and are called automatically – they cannot be called explicitly.

Values to initialize class variables are passed to the constructor method in the statement creating an object, in parentheses following the object name.

constructor.cpp

Hot tip

The definition of a class method is also known as the method "implementation".

1 Rename a copy of the previous example "multiple.cpp" as a new program "constructor.cpp"

2 In the public section of the Dog class declaration, replace the setValues method prototype with this constructor prototype
Dog (int, int, string) ;

3 Now, add an associated destructor prototype
~Dog() ;

4 After the Dog class declaration, replace the setValues definition block with a constructor definition block
Dog::Dog (int age, int weight, string color)
{

}

5 In the constructor definition block, insert three statements to assign values from passed arguments to class variables
this -> age = age ;
this -> weight = weight ;
this -> color = color ;

6 After the constructor definition, add a destructor definition block

Dog::~Dog()
{

}

7 In the destructor definition, insert a statement to output a confirmation whenever an instance object gets destroyed

cout << "Object destroyed." << endl ;

8 In the main method, edit the statement creating the "fido" object – to pass values to its constructor method

Dog fido(3, 15, "brown") ;

9 Similarly, edit the statement creating the "pooch" object – to pass values to the constructor method

Dog pooch(4, 18, "gray") ;

10 Delete the statements calling the setValues method of the "fido" and "pooch" objects – the constructor has now replaced that method

11 Save, compile, and run the program – see the output appear as before, plus confirmation when the objects get destroyed

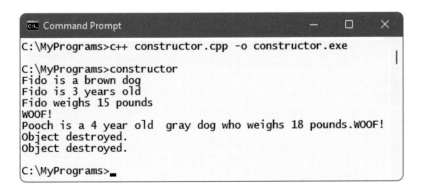

```
Command Prompt                                    —    □    ×

C:\MyPrograms>c++ constructor.cpp -o constructor.exe

C:\MyPrograms>constructor
Fido is a brown dog
Fido is 3 years old
Fido weighs 15 pounds
WOOF!
Pooch is a 4 year old  gray dog who weighs 18 pounds.WOOF!
Object destroyed.
Object destroyed.

C:\MyPrograms>_
```

Beware

The destructor definition begins with the class name "Dog", the scope resolution operator "::", then the destructor method name "~Dog".

Don't forget

Although the initial values of the variable members are set by the constructor, setter methods can be added to subsequently change the values – and those new values can be retrieved by the getter methods.

Overloading methods

Just as C++ allows functions to be overloaded, class methods can be overloaded too – including constructor methods. An overloaded constructor method is useful to assign default values to member variables when an object is created without passing values to the constructor.

overloaded.cpp

1 Rename a copy of the previous example "constructor.cpp" as a new program "overloaded.cpp"

2 In the public section of the Dog class declaration, add inline an overloaded bark method – to output a passed string argument when called
void bark (string noise) { cout << noise << endl ; }

3 Now, declare a constructor method prototype that takes no arguments (a default constructor method) and an overloaded constructor method prototype that takes two arguments
Dog() ;
Dog (int, int) ;

4 After the Dog class declaration, add a definition for the default constructor method – assigning default values to class variables when an object is created without passing any arguments
Dog::Dog()
{
    age = 1 ;
    weight = 2 ;
    color = "black" ;
}

Don't forget

The **this ->** pointer is used to explicitly identify class members when arguments have the same name as members.

5 Now, add a definition for the overloaded constructor method – assigning default values to class variables when an object is created passing two arguments
Dog::Dog (int age, int weight)
{
    this -> age = age ;
    this -> weight = weight ;
    color = "white" ;
}

6 In the main method, insert a statement to create a Dog object without passing any arguments – calling the default constructor
Dog rex ;

Beware

7 Add statements calling each getter method to retrieve the default values – set by the default constructor method
cout << "Rex is a " << rex.getAge() ;
cout << " year old " << rex.getColor() ;
cout << " dog who weighs " << rex.getWeight() ;
cout << " pounds ." ;

Don't add parentheses after the object name when creating an object without passing arguments – notice it's
Dog rex ; not
Dog rex() ;

8 Now, add a call to the overloaded output method
rex.bark("GRRR!") ;

9 Insert a statement to create a Dog object passing two arguments – to call the overloaded constructor
Dog sammy(2, 6) ;

10 Add statements to retrieve the values set by the overloaded constructor method and call the overloaded output method
cout << "Sammy is a " << sammy.getAge() ;
cout << " year old " << sammy.getColor() ;
cout << " dog who weighs " << sammy.getWeight() ;
cout << " pounds ." ;
sammy.bark("BOWOW!") ;

Hot tip

This is the final rendition of the Dog class. Be sure you can readily identify its public and private members before proceeding.

11 Save, compile, and run the program

```
Command Prompt                           —    □    ×

C:\MyPrograms>c++ overloaded.cpp -o overloaded.exe

C:\MyPrograms>overloaded
Fido is a brown dog
Fido is 3 years old
Fido weighs 15 pounds
WOOF!
Pooch is a 4 year old gray dog who weighs 18 pounds.WOOF!
Rex is a 1 year old black dog who weighs 2 pounds.GRRR!
Sammy is a 2 year old white dog who weighs 6 pounds.BOWOW!
Object destroyed.
Object destroyed.
Object destroyed.
Object destroyed.

C:\MyPrograms>_
```

Inheriting class properties

A C++ class can be created as a brand new class, like those in previous examples, or can be "derived" from an existing class. Importantly, a derived class inherits members of the parent (base) class from which it is derived – in addition to its own members.

The ability to inherit members from a base class allows derived classes to be created that share certain common properties, which have been defined in the base class. For example, a "Polygon" base class may define width and height properties that are common to all polygons. Classes of "Rectangle" and Triangle" could be derived from the Polygon class – inheriting width and height properties, in addition to their own members defining their unique features.

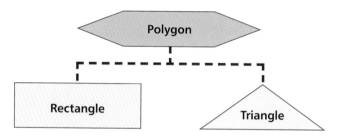

The virtue of inheritance is extremely powerful and is the second principle of Object Oriented Programming (OOP).

A derived class declaration adds a colon : after its class name, followed by an access specifier and the class from which it derives.

derived.cpp

1 Start a new program by specifying the C++ library classes to include, and a namespace prefix to use
```
#include <iostream>
using namespace std ;
```

2 Declare a class named "Polygon" containing two protected variables, accessible only to members of this class and classes derived from this class, along with a public method to assign values to those variables
```
class Polygon
{
  protected:
    int width, height ;
  public:
    void setValues( int w, int h ) { width = w; height = h ; }
} ;
```

3 After the Polygon class, declare a Rectangle class that derives from the Polygon class and adds a unique method

```
class Rectangle : public Polygon
{
  public:
    int area() { return ( width * height ) ; }
} ;
```

4 After the Rectangle class, declare a Triangle class that derives from the Polygon class and adds a unique method

```
class Triangle : public Polygon
{
  public:
    int area() { return ( ( width * height ) / 2 ) ; }
} ;
```

5 After the Triangle class, add a main method containing a final **return** statement and creating an instance of each derived class

```
int main()
{
  Rectangle rect ; Triangle trgl ;
  return 0 ;
}
```

6 Insert calls to the method inherited from the Polygon base class – to initialize the inherited variables

```
rect.setValues( 4, 5 ) ;
trgl.setValues( 4, 5 ) ;
```

7 Output the value returned by the unique method of each derived class

```
cout << "Rectangle area : " << rect.area() << endl ;
cout << "Triangle area : " << trgl.area() << endl ;
```

8 Save, compile, and run the program to see the output

```
C:\MyPrograms>c++ derived.cpp -o derived.exe

C:\MyPrograms>derived
Rectangle area : 20
Triangle area : 10

C:\MyPrograms>
```

Beware

Don't confuse class instances and derived classes – an instance is a copy of a class, whereas a derived class is a new class that inherits properties of the base class from which it is derived.

Hot tip

A class declaration can derive from more than one class – for example, **class Box : public A, public B, public C { } ;**

127

Calling base constructors

Although derived classes inherit the members of their parent base class, they do not inherit its constructor and destructor. However, it should be noted that the default constructor of the base class is always called when a new object of a derived class is created – and the base class destructor is called when the object gets destroyed. These calls are made in addition to those made to the constructor and destructor methods of the derived class.

The default constructor of the base class has no arguments – but that class may also have overloaded constructors that do. If you prefer to call an overloaded constructor of the base class when a new object of a derived class is created, you can create a matching overloaded constructor in the derived class – having the same number and type of arguments.

basecon.cpp

1 Start a new program by specifying the C++ library classes to include, and a namespace prefix to use
```
#include <iostream>
using namespace std ;
```

2 Declare a class named "Parent", which will be a base class
```
class Parent
{
  // Class members go here.
} ;
```

3 Between the braces of the Parent class declaration, insert a public access specifier and add a default constructor to output text – identifying when it has been called
```
public:
  Parent()
  { cout << "Default Parent constructor called." ; }
```

Son – Parent – Daughter

4 Add an overloaded constructor, which takes a single integer argument and also outputs identifying text
```
  Parent ( int a )
  { cout << endl <<
        "Overloaded Parent constructor called." ; }
```

5 After the Parent class, declare a derived "Daughter" class
```
class Daughter : public Parent
{

} ;
```

6 In the Daughter class declaration, insert a public access specifier and add a default constructor to output text – identifying when it has been called

```
public :
  Daughter ()
  { cout << endl <<
"    Derived Daughter class default constructor called." ; }
```

Notice that the syntax in the overloaded Son class constructor passes the integer argument to the overloaded base class constructor.

7 After the Daughter class, declare a derived "Son" class

```
class Son : public Parent
{

} ;
```

8 In the Son class declaration, insert a public access specifier and add an overloaded constructor that takes a single integer argument, and also outputs identifying text

```
public :
  Son ( int a ) : Parent ( a )
  { cout << endl <<
"    Derived Son class overloaded constructor called." ; }
```

Each class automatically has an empty default constructor and destructor – for example, **Son( ){ }** and **~Son( ){ }**

9 After the Son class, add a main method containing a final **return** statement and creating an instance of each derived class – calling base class and derived class constructors

```
int main()
{
  Daughter emma ;
  Son andrew(0) ;
  return 0 ;
}
```

10 Save, compile, and run the program to see the output from each constructor in turn as it gets called

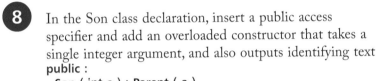

```
Command Prompt                          —    □    ×

C:\MyPrograms>c++ basecon.cpp -o basecon.exe

C:\MyPrograms>basecon
Default Parent constructor called.
    Derived Daughter class default constructor called.
Overloaded Parent constructor called.
    Derived Son class overloaded constructor called.

C:\MyPrograms>_
```

Overriding base methods

A method can be declared in a derived class to override a matching method in the base class – if both method declarations have the same name, arguments, and return type. This effectively hides the base class method as it becomes inaccessible unless it is called explicitly, using the :: scope resolution operator for precise identification.

The technique of overriding base class methods must be used with care, however, to avoid unintentionally hiding overloaded methods – a single overriding method in a derived class will hide <u>all</u> overloaded methods of that name in the base class!

override.cpp

Don't forget

The method declaration in the derived class must exactly match that in the base class to override it – including the **const** keyword if it is used.

1 Start a new program by specifying the C++ library classes to include, and a namespace prefix to use
```
#include <string>
#include <iostream>
using namespace std ;
```

2 Declare a class named "Man", which will be a base class
```
class Man
{
  // Class members go here.
} ;
```

3 Between the braces of the Man class declaration, insert a public access specifier and an inline output method
```
public :
  void speak() { cout << "Hello! " << endl ; }
```

4 Now, insert an overloaded inline output method
```
  void speak( string msg )
  { cout << "    " << msg << endl ; }
```

5 After the Man class declaration, declare a class named "Hombre" that is derived from the Man class
```
class Hombre : public Man
{
  // Class members go here.
} ;
```

6 Between the braces of the Hombre class declaration, insert an access specifier and a method that overrides the overloaded base class method – without a tab output

```
public :
  void speak( string msg ) { cout << msg << endl ; }
```

7 After the Hombre class declaration, add a main method containing a final **return** statement and creating two objects – an instance of the base class and an instance of the derived class

```
int main()
{
  Man henry ;
  Hombre enrique ;
  // Add more statements here.
  return 0 ;
}
```

Henry   Enrique

8 In the main method, insert statements calling both methods of the base class

```
henry.speak() ;
henry.speak( "It's a beautiful evening." ) ;
```

9 Next, insert a statement calling the overriding method in the derived class – producing output without a tab

```
enrique.speak( "Hola!" ) ;
```

10 Now, insert a statement explicitly calling the overridden method in the base class

```
enrique.Man::speak( "Es una tarde hermosa." ) ;
```

11 Save, compile, and run the program to see the output from the overriding and overridden methods

The overriding method declared in the derived class hides both overloaded classes in the base class. Try calling **enrique.speak()** – the compiler will complain there is no matching method for that call.

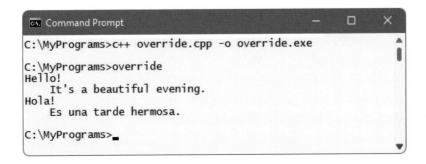

```
Command Prompt                                    —   □   ×

C:\MyPrograms>c++ override.cpp -o override.exe

C:\MyPrograms>override
Hello!
    It's a beautiful evening.
Hola!
    Es una tarde hermosa.

C:\MyPrograms>_
```

Summary

- The first principle of Object Oriented Programming is the encapsulation of data and functionality within a single class.

- Access specifiers **public, private**, and **protected** control the accessibility of class members from outside the class.

- A class declaration describes a data structure from which instance objects can be created.

- Public setter and getter class methods store and retrieve data from private class variable members.

- The scope resolution operator :: can explicitly identify a class.

- Class members that have the same name as a passed argument can be explicitly identified by the **this ->** pointer.

- A constructor method is called when an object gets created, and a destructor method is called when it gets destroyed.

- Class variables can be automatically initialized by a constructor.

- Class methods can be overloaded like other functions.

- The second principle of Object Oriented Programming is the virtue of inheritance that allows derived classes to inherit the properties of their parent base class.

- In a derived class declaration, the class name is followed by a : colon character, an access specifier, and its base class name.

- When an instance object of a derived class is created, the default constructor of the base class gets called in addition to the constructor of the derived class.

- A derived class method can override a matching method in its base class – also overriding all overloaded methods of that name within the base class.

8 Harnessing polymorphism

Pointing to classes

The three cornerstones of Object Oriented Programming (OOP) are encapsulation, inheritance, and polymorphism. Examples in Chapter 7 have demonstrated how data can be encapsulated within a C++ class, and how derived classes inherit the properties of their base class. This chapter introduces the final cornerstone principle of polymorphism.

The term "polymorphism" (from Greek, meaning "many forms") describes the ability to assign a different meaning, or purpose, to an entity according to its context.

In C++, overloaded operators can be described as polymorphic. For example, the * character can represent either the multiply operator or the dereference operator, according to its context. Similarly, the + character can represent either the add operator or the concatenate operator, according to its context.

More importantly, C++ class methods can also be polymorphic. The key to understanding polymorphism with classes is to recognize that a base class pointer can be created that is also bound to a particular derived class by association.

Turn back to Chapter 6 for more on pointers.

A pointer to a base class can be assigned the memory address of a derived class to provide a "context" – to uniquely identify that derived class. For example, with a **Base** base class and a derived **Sub** class, a pointer can be created like this:

```
Sub inst ;
Base* pSub = &inst ;
```

or more simply using the **new** keyword, like this:

```
Base* pSub = new Sub ;
```

Where there are multiple derived classes, base class pointers can be created binding each derived class by its unique memory address – which can be revealed using the addressof **&** operator.

classptr.cpp

1 Start a new program by specifying the C++ library classes to include, and a namespace prefix to use
```
#include <iostream>
using namespace std ;
```

2 Declare a **Base** class containing a method to output a passed integer value in hexadecimal format

```
class Base
{
  public:
  void Identify( int adr ) const
  {
    cout << "Base class called by 0x"
                << hex << adr << endl ; }
} ;
```

The **->** class pointer operator is used here to call class methods.

3 Declare two empty derived classes, **SubA** and **SubB**

```
class SubA : public Base {        } ;
class SubB : public Base {        } ;
```

4 Declare a main method containing a final **return** statement

```
int main()
{
  // Program code goes here.
  return 0 ;
}
```

5 In the main method, insert statements to create two base class pointers – each binding to a specific derived class

```
Base* ptrA = new SubA ;
Base* ptrB = new SubB ;
```

The hexadecimal address is passed as an **int** data type, then displayed in hexadecimal format by the **hex** output manipulator. The addresses will be different each time the program executes – they are assigned dynamically.

6 Now, insert statements that use the pointers to call the base class method, passing the memory address of each for output

```
ptrA -> Identify( (int) &ptrA ) ;
ptrB -> Identify( (int) &ptrB ) ;
```

7 Save, compile, and run the program to see the addresses

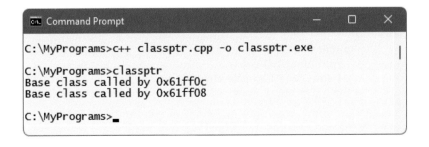

```
C:\MyPrograms>c++ classptr.cpp -o classptr.exe

C:\MyPrograms>classptr
Base class called by 0x61ff0c
Base class called by 0x61ff08

C:\MyPrograms>
```

Calling a virtual method

A base class pointer that is bound to a specific derived class can be used to call derived class methods that have been inherited from the base class. Methods that are unique to the derived class must, however, be called via an instance object.

A base class pointer that is bound to a specific derived class can also be used to explicitly call a method in the base class using the :: scope resolution operator.

Most usefully, an inherited method in a derived class can override that in the base class when the base method has been declared as a "virtual" method. This is just a regular method declaration in the base class preceded by the **virtual** keyword. The declaration of a virtual method indicates that the class will be used as a base class from which another class will be derived, which may contain a method to override the virtual base method.

virtual.cpp

Pointers to a base class cannot be used to call non-inherited methods in a derived class.

1. Start a new program by specifying the C++ library classes to include, and a namespace prefix to use
```
#include <iostream>
using namespace std ;
```

2. Declare a base class named "Parent", containing a regular method declaration and a virtual method declaration
```
class Parent
{
  public :
    void Common() const
    { cout << "I am part of this family, " ; }

    virtual void Identify() const
    { cout << "I am the parent" << endl ; }
} ;
```

3. Declare a derived class named "Child", containing a method to override the virtual base method
```
class Child : public Parent
{
  public :
    void Identify() const
    { cout << "I am the child" << endl ; }
} ;
```

...cont'd

4 Declare a "Grandchild" class, derived from the "Child" class, containing a method to override the virtual base method and a regular method declaration

```cpp
class Grandchild : public Child
{
  public :
    void Identify() const
    { cout << "I am the grandchild" << endl ; }

    void Relate() const
    { cout << "Grandchild has parent and grandparent" ; }
} ;
```

5 Declare a main method containing a final **return** statement and creating instances of each derived class, plus base class pointers binding those derived classes

```cpp
int main()
{
  Child son ;
  Grandchild grandson ;
  Parent* ptrChild = &son ;
  Parent* ptrGrandchild = &grandson ;
  // Add more statements here.
  return 0 ;
}
```

Parent
Child
Grandchild

137

6 In the main method, insert calls to each method

```cpp
ptrChild -> Common() ;            // Inherited.
ptrChild -> Identify() ;          // Overriding.
ptrGrandchild -> Common() ;       // Inherited.
ptrGrandchild -> Identify() ;     // Overriding.
ptrChild -> Parent::Common() ;    // Explicit.
ptrChild -> Parent::Identify() ;  // Explicit.
grandson.Relate() ;               // Via instance.
```

Don't forget

Here, the Grandchild class inherits the properties of the Child class, which inherits the properties of the Parent class.

7 Save, compile, and run the program to see the output

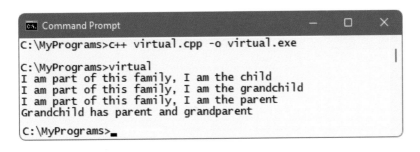

```
Command Prompt                        —    □    ✕

C:\MyPrograms>c++ virtual.cpp -o virtual.exe

C:\MyPrograms>virtual
I am part of this family, I am the child
I am part of this family, I am the grandchild
I am part of this family, I am the parent
Grandchild has parent and grandparent

C:\MyPrograms>_
```

Directing method calls

The great advantage of polymorphism with multiple derived class objects is that calls to methods of the same name are directed to the appropriate overriding method.

A base class may contain only virtual methods that each derived class may override with their own methods, but base class methods can still be called explicitly using the :: scope resolution operator. This can allow inconsistencies, however – this example would seem to imply that chickens can fly!

birds.cpp

1. Start a new program by specifying the C++ library classes to include, and a namespace prefix to use
```
#include <iostream>
using namespace std ;
```

2. Declare a base class named "Bird", containing two virtual method declarations
```
class Bird
{
  public :
    virtual void Talk() const
    { cout << "A bird talks... " << endl ; }

    virtual void Fly() const
    { cout << "A bird flies... " << endl ; }
} ;
```

Hot tip

Overriding methods in a derived class may, optionally, include the **virtual** prefix – as a reminder it is overriding a base class method.

3. Declare a derived class named "Pigeon", containing two methods to override those in the base class
```
class Pigeon : public Bird
{
  public :
    void Talk() const
    { cout << "Coo! Coo!" << endl ; }

    void Fly() const
    { cout << "A pigeon flies away... " << endl ; }
} ;
```

4 Declare a derived class named "Chicken", containing two methods to override those in the base class

```cpp
class Chicken : public Bird
{
  public :
    void Talk() const
    { cout << "Cluck! Cluck!" << endl ; }

    void Fly() const
    { cout << "I\'m just a chicken – I can\'t fly!" << endl ; }
} ;
```

Chicken   Pigeon
(Bird)   (Bird)

5 Declare a main method containing a final **return** statement and creating base class pointers binding derived classes

```cpp
int main()
{
  Bird* pPigeon = new Pigeon ;
  Bird* pChicken = new Chicken ;
  // Add more statements here.
  return 0 ;
}
```

6 In the main method, insert calls to each method

```cpp
pPigeon -> Talk() ;
pPigeon -> Fly() ;
pChicken -> Talk() ;
pChicken -> Fly() ;
pPigeon -> Bird::Talk() ;
pChicken -> Bird::Fly() ;            // Inappropriate call.
```

The backslash \ character is required to escape the apostrophe in strings.

7 Save, compile, and run the program to see the output

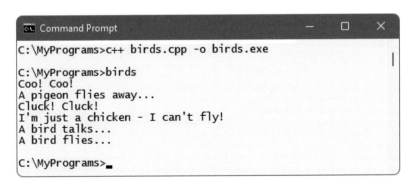

```
C:\MyPrograms>c++ birds.cpp -o birds.exe

C:\MyPrograms>birds
Coo! Coo!
A pigeon flies away...
Cluck! Cluck!
I'm just a chicken - I can't fly!
A bird talks...
A bird flies...

C:\MyPrograms>
```

Providing capability classes

Classes whose sole purpose is to allow other classes to be derived from them are known as "capability classes" – they provide capabilities to the derived classes.

Capability classes generally contain no data, but merely declare a number of virtual methods that can be overridden in their derived classes.

The following example builds upon the previous example to demonstrate how the "Bird" class can be better written as a capability class. Its methods no longer contain output statements, but return a **-1** (error) value if they are called explicitly.

It is necessary to change the return type of those methods from **void** to **int**, and these changes must also be reflected in each overriding method in the derived classes.

capability.cpp

1 Start a new program by specifying the C++ library classes to include, and a namespace prefix to use
```
#include <iostream>
using namespace std ;
```

2 Declare a base capability class named "Bird", containing two virtual method declarations that will signal an error if called explicitly
```
class Bird
{
  public :
    virtual int Talk() const { return -1 ; }
    virtual int Fly()  const { return -1 ; }
} ;
```

The return value of overriding methods in derived classes must match those declared in the base class.

3 Declare a derived class named "Pigeon", containing two methods to override those in the base class
```
class Pigeon : public Bird
{
  public :
    int Talk() const
    { cout << "Coo! Coo!" << endl ; return 0 ; }

    int Fly() const
    { cout << "A pigeon flies away..." << endl ; return 0 ; }
} ;
```

4 Declare a derived class named "Chicken" containing two methods to override those in the base class

```
class Chicken : public Bird
{
  public :
    int Talk() const
    { cout << "Cluck! Cluck!" << endl ; return 0 ; }

    int Fly() const
    { cout << "I\'m just a chicken – I can\'t fly!"
                            << endl ; return 0 ; }
} ;
```

Beware

Capability class methods are intended to be overridden in derived classes – they should not be called explicitly.

5 Declare a main method creating base class pointers binding the derived classes

```
int main()
{
  Bird* pPigeon = new Pigeon ;
  Bird* pChicken = new Chicken ;
}
```

6 In the main method, insert method calls and a statement to terminate the program when an error is met by explicitly calling a base class method

```
pPigeon -> Talk() ;
pChicken -> Talk() ;

pPigeon -> Bird::Talk() ;
if ( -1 ) { cout << "Error! - Program ended."
                            << endl ; return 0 ; }

pPigeon -> Fly() ;      // Call will not be made.
pChicken -> Fly() ;     // Call will not be made.
return 0 ;              // Statement will not be executed.
```

Hot tip

Refer back to pages 90-95 for more details on error handling.

141

7 Save, compile, and run the program to see the output

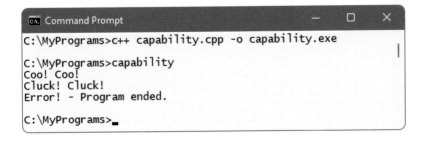

```
C:\MyPrograms>c++ capability.cpp -o capability.exe

C:\MyPrograms>capability
Coo! Coo!
Cluck! Cluck!
Error! - Program ended.

C:\MyPrograms>
```

Making abstract data types

An Abstract Data Type (ADT) represents a concept, rather than a tangible object, and is always the base to other classes. A base class can be made into an ADT by initializing one or more of its methods with 0. These are known as "pure virtual methods" and must always be overridden in derived classes.

adt.cpp

1 Start a new program by specifying the C++ library classes to include, and a namespace prefix to use

```
#include <iostream>
using namespace std ;
```

2 Declare a base ADT class named "Shape", containing three pure virtual methods

```
class Shape
{
  public :
    virtual int getArea() = 0 ;
    virtual int getEdge() = 0 ;
    virtual void Draw() = 0 ;
} ;
```

Beware

It is illegal to create an instance object of an ADT – attempting to do so will simply create a compiler error.

3 Declare a derived class named "Rect", containing two private variables

```
class Rect : public Shape
{
  private :
    int height, width ;
} ;
```

4 In the derived class declaration, insert a public constructor and destructor

```
public :
  Rect( int initWidth, int initHeight )
  {
    height = initHeight ;
    width = initWidth ;
  }

  ~Rect() ;
```

5 In the derived "Rect" class declaration, declare three public methods to override the pure virtual methods declared in the "Shape" base class

```
int getArea() { return height * width } ;
int getEdge() { return ( 2 * height ) + ( 2 * width ) ; }

void Draw()
{
  for ( int i = 0 ; i < height ; i++ ) {
    for ( int j = 0 ; j < width ; j++ ) { cout << "x " ; }
  cout << endl ; }
}
```

6 Declare a main method containing a final **return** statement and creating two instances of the derived "Rect" class – to represent a Square and a Quadrilateral shape

```
int main
{
  Shape* pQuad = new Rect( 3, 7 ) ;
  Shape* pSquare = new Rect( 5, 5 ) ;
  // Add more statements here.
  return 0 ;
}
```

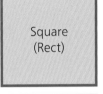

Square
(Rect)

Quadrilateral
(Rect)

7 In the main method, insert calls to each method then save, compile, and run the program to see the output

```
pQuad -> Draw() ;
cout << "Area is " << pQuad -> getArea() << endl ;
cout << "Perimeter is " << pQuad -> getEdge() << endl ;

pSquare -> Draw() ;
cout << "Area is " << pSquare -> getArea() << endl ;
cout << "Perimeter is "<< pSquare -> getEdge() <<endl ;
```

A base class need only contain one pure virtual method to create an Abstract Data Type.

```
Command Prompt                                    —  □  ✕

C:\MyPrograms>c++ adt.cpp -o adt.exe

C:\MyPrograms>adt
x x x x x x x
x x x x x x x
x x x x x x x
Area is 21
Perimeter is 20
x x x x x
x x x x x
x x x x x
x x x x x
x x x x x
Area is 25
Perimeter is 20
```

Building complex hierarchies

It is sometimes desirable to derive an ADT from another ADT to construct a complex hierarchy of classes. This provides great flexibility and is perfectly acceptable, providing each pure method is defined at some point in a derived class.

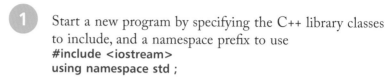
hierarchy.cpp

1 Start a new program by specifying the C++ library classes to include, and a namespace prefix to use
```
#include <iostream>
using namespace std ;
```

2 Declare a base ADT class named "Boat", containing a variable and accessor method together with one pure virtual method
```
class Boat
{
  protected:
    int length ;
  public :
    int getLength() { return length ; }
    virtual void Model() = 0 ;
} ;
```

The Boat class has properties common to any boat, whereas the Sailboat class has properties specific to boats that have sails.

3 Declare an ADT class (derived from the Boat class) named "Sailboat" – also containing a variable and accessor method together with one pure virtual method
```
class Sailboat : public Boat
{
  protected :
    int mast ;
  public :
    int getMast() { return mast ; }
    virtual void Boom() = 0 ;
} ;
```

Laser
Sailboat (Boat)

4 Declare a regular class (derived from the Sailboat class) named "Laser", in which all members will allow public access
```
class Laser : public Sailboat
{
  public :

} ;
```

5 In the Laser class, insert a call to its constructor method to assign values to the variables in each class from which this class is derived – and call the destructor method
Laser() { mast = 19 ; length = 35 ; }
~Laser() ;

6 In the Laser class, define the pure virtual methods declared in each class from which this class is derived
void Model() { cout << "Laser Classic" << endl ; }
void Boom() { cout << "Boom: 14ft" << endl ; }

7 Declare a main method containing a final **return** statement and creating an instance of the derived class on the bottom tier of the hierarchy
```
int main()
{
  Laser* pLaser = new Laser ;
  // Add more statements here.
  return 0 ;
}
```

Hot tip

Try adding a Powerboat class derived from the Boat class (to contain engine information), and a Cruiser class derived from the Powerboat class – to assign variable values and to define virtual methods.

8 In the main method, insert calls to each defined method
```
pLaser -> Model() ;
cout << "Length: " <<
        pLaser -> getLength() << "ft" << endl ;
cout << "Height: "<<
        pLaser -> getMast() << "ft" << endl ;
pLaser -> Boom() ;
```

9 Save, compile, and run the program to see the output

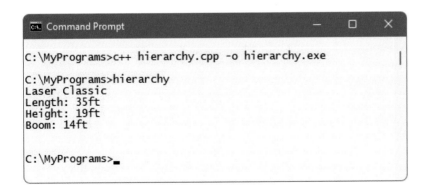

```
C:\MyPrograms>c++ hierarchy.cpp -o hierarchy.exe

C:\MyPrograms>hierarchy
Laser Classic
Length: 35ft
Height: 19ft
Boom: 14ft


C:\MyPrograms>
```

Isolating class structures

The source code for each example program in this book is generally contained in a single **.cpp** file to save space, but in reality OOP programs are often contained in three separate files:

- Header **.h** file – contains only class declarations.

- Implementation **.cpp** file – contains class definitions to implement the methods declared in the header file, which is referenced by an **#include** directive.

- Client **.cpp** file – contains a **main** method that employs the class members declared in the header file, which is also referenced by an **#include** directive.

For example, a sum calculator program might comprise three files named **ops.h** (a header file declaring operation classes), **ops.cpp** (an implementation file defining the operation methods), and **sum.cpp** (a client file calling the various operations).

When compiling **sum.cpp**, the compiler incorporates the included header file and implementation file into the program. It first translates the header file and implementation file into a binary object file (**ops.o**), then it translates the header file and client file into a second binary object file (**sum.o**). Finally, the Linker combines both object files into a single executable file (**sum.exe**).

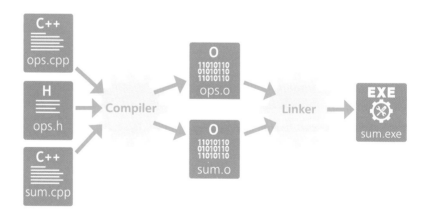

Isolating class structures in separate "modular" files has the advantage of improving portability, and makes code maintenance easier by clearly identifying the purpose of each file.

To have the compiler combine all three source code files into a single, executable file it is only necessary to explicitly specify the **.cpp** files in the compiler command – an **#include** directive ensures the header file will also be recognized.

For example, with the statement **#include "ops.h"** in both **ops.cpp** and **sum.cpp**, the command to compile the example described opposite need not specify the **ops.h** header in the compiler command, but is just **c++ ops.cpp sum.cpp -o sum.exe**.

This example will allow the user to input a simple arithmetical expression and output the calculated result. It will provide instructions when first launched and allow the user to make subsequent expressions – or exit by entering a 0 character.

Calculator

ops.h

1. Start a header file by declaring a class named "Calculator"
   ```
   class Calculator
   {

   } ;
   ```

2. In the class declaration, insert public method declarations
   ```
   public :
     Calculator() ;          // (Constructor) To set initial status.
     void launch() ;         // To display initial instructions.
     void readInput() ;      // To get expression.
     void writeOutput() ;    // To display result.
     bool run() ;            // (Accessor) To get current status.
   ```

3. In the class declaration, insert private variable declarations
   ```
   private :
     double num1, num2 ;     // To store input numbers.
     char oper ;             // To store input operator.
     bool status ;           // To store current status.
   ```

Beware

4. Save the header file as "ops.h"

5. Turn to page 148 to continue this example by creating an implementation file – containing definitions for the Calculator class methods declared in the "ops.h" header file

Notice that the header file name must be surrounded by quotes in an **#include** directive – not by the < > angled brackets used to include a standard C++ library.

Employing isolated classes

ops.cpp

6 Start an implementation file with **include** directives for the header file created on page 147, and the standard C++ library supporting input/output statements

```cpp
#include "ops.h"              // Reference header file.
#include <iostream>
using namespace std ;
```

7 Add the following definitions for each method in the header file, then save the implementation file as "ops.cpp"

```cpp
Calculator::Calculator()
{ status = true ; }          // Initialize status.

void Calculator::launch()    // Display instructions.
{
  cout << endl << "*** SUM CALCULATOR ***" << endl ;
  cout << "Enter a number, an operator(+,-,*,/), and
            another number." << endl << "Hit Return to
            calculate. Enter 0 to exit." << endl ;
}

void Calculator::readInput()             // Get expression.
{
  cout << "> " ;  cin >> num1 ;          // Get 1st number.
  if ( num1 == 0 ) status = false ;      // Exit if it's 0.
  else { cin >> oper ; cin >> num2 ; }   // Or get the rest.
}

void Calculator::writeOutput()           // Display result.
{
  if ( status ) switch( oper )           // If continuing.
  {                                      // Show result.
  case '+' : { cout << ( num1 + num2 ) << endl ; break ; }
  case '-' : { cout << ( num1 - num2 ) << endl ; break ; }
  case '*' : { cout << ( num1 * num2 ) << endl ; break ; }
  case '/' : if ( num2 != 0 )
          cout << ( num1 / num2 ) << endl ;
          else cout << "Cannot divide by 0" << endl ;
  }
}

bool Calculator::run()       // Get the current status.
{ return status ; }
```

Beware

Due to space limitation, this program makes barely any attempt at input validation – it assumes the user will enter a valid expression, such as 8 * 3.

8   Start a client file with an include directive to incorporate the header file created on page 147
```cpp
#include "ops.h"
```

sum.cpp

9   Declare a main method containing a final **return** statement, and creating a pointer plus a call to display instructions
```cpp
int main()
{
  Calculator* pCalc = new Calculator ;
  pCalc -> launch() ;
  // Add more statements here.
  return 0 ;
}
```

10 In the main method, insert a loop to read expressions and write results while the program status permits
```cpp
while ( pCalc -> run() )
{
  pCalc -> readInput() ;
  pCalc -> writeOutput() ;
}
```

11 Save the client file as "sum.cpp", alongside "ops.h" and "ops.cpp", then compile the program with this command
```
c++ ops.cpp sum.cpp -o sum.exe
```

12 Run the program and enter simple expressions to see the results, then enter 0 and hit **Return** to exit the program

```
Command Prompt                          —    □    ×
C:\MyPrograms>c++ ops.cpp sum.cpp -o sum.exe
C:\MyPrograms>sum                                    |

*** SUM CALCULATOR ***
Enter a number, an operator(+,-,*,/), and another number.
Hit Return to calculate. Enter zero to exit.
> 32 + 32
64
> 5.25 - 8.75
-3.5
> 8 * 3
24
> 20 / 5
4
> 20 / 0
Cannot divide by zero
> 0

C:\MyPrograms>_
```

149

This program loops until the user types a 0 and hits **Return** – changing the "status" control variable to **false,** and so exiting the program.

Summary

- The three cornerstones of Object Oriented Programming are encapsulation, inheritance, and polymorphism.

- Polymorphic entities have a different meaning, or purpose, according to their context.

- A base class pointer can be used to call inherited methods in the derived class to which it is bound.

- A base class pointer can also be used to explicitly call base class methods using the :: scope resolution operator.

- Virtual base class methods are intended to be overridden in derived classes.

- Polymorphism allows calls to methods of the same name to be directed to the appropriate overriding method.

- Capability classes generally contain no data, but merely declare virtual methods that can be overridden in derived classes.

- Virtual methods that return a **-1** value signal an error to indicate they should not be called directly.

- An Abstract Data Type represents a concept, and is always the base to other classes.

- Declaration of a pure virtual method, with the assignation **=0**, indicates that class is an ADT.

- Classes can be derived from an ADT – but you cannot create an instance of an ADT.

- An ADT can be derived from another ADT to create a complex hierarchy of classes.

- Programs can be separated into header, implementation, and client files to aid portability and to ease code maintenance.

- Header files that are referenced by **#include** directives will be automatically included by the compiler during compilation.

9 Processing macros

This chapter demonstrates how the C++ compiler can be made to perform useful tasks before compiling a program.

Exploring compilation

Whenever the C++ compiler runs, it first calls upon its preprocessor to seek any compiler directives that may be included in the source code. Each of these begin with the **#** hash character and will be implemented first to effectively modify the source code before it is assembled and compiled.

The changes made by compiler directives to the preprocessor create new temporary files that are not normally seen. It is these temporary files that are used to create a binary object file:

Source Code (.cpp)

Preprocessor

Substitutions (.ii)

Compiler

Assembly Code (.s)

Assembler

Object Code (.o)

Linker

Executable (.exe)

- The first temporary file created during compilation expands the original source code by replacing its compiler directives with library code that implements those directives. This text file is named like the source file, but with a **.ii** file extension.

- The second temporary file created during compilation is a translation of the temporary expanded **.ii** file into low-level Assembly language instructions. This text file is named like the source file, but with a **.s** file extension.

- The third temporary file created during compilation is a translation of the temporary Assembly language **.s** file into machine code. This binary object file is named like the source file, but with a **.o** file extension.

So, the compilation process employs the Preprocessor to compile source code, an "Assembler" to translate this into machine code, and a Linker to convert one or more binary objects into an executable program.

You can see the temporary files by instructing the compiler to save them using the **-save-temps** compiler option. Both temporary text files can then be examined by opening them in a plain text editor.

Most significantly, you can see that the temporary file with the **.ii** file extension contains the complete function definitions from any included library. For example, it replaces an **#include <iostream>** directive with definitions for the **cin, cout, cerr** functions, and the **clog** function that can be used to redirect error messages to a file. The end of the **.ii** file shows the defined functions to be part of the "std" namespace – so they can appear without the **std::** prefix.

...cont'd

1 Create a simple program named "prog.cpp" that will
output a message when it gets executed

prog.cpp

```cpp
#include <iostream>
using namespace std ;

int main()
{
  cout << "This is a simple test program" << endl ;
  return 0 ;
}
```

2 Issue a command using the **-save-temps** option, to save
temporary files, and a **-c** option to compile this program's
source files into an object file – with no executable file

c++ prog.cpp -save-temps -c

One or more object files
can be used to create
an executable file – as
described on page 146.

3 Open
the **.ii**
file in
a plain
text
editor
such as
Notepad,
then

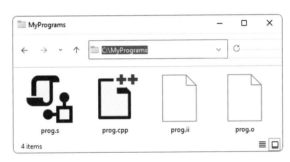

scroll to the end of the file to see the modified source
code – notice how the **<iostream>** library functions are
defined in the **std** namespace

153

4 Open the **.s** file in a plain text editor to see the low-level
assembler instructions – notice how the message string is
now terminated by the special **\0** character

You can combine
these steps, creating
an executable file and
saving temporary files,
by issuing the command
**c++ prog.cpp
-save-temps -o prog.exe**

5 Issue a command to output an executable file from the
.o object file, then run the program to see the message

c++ prog.o -o prog.exe

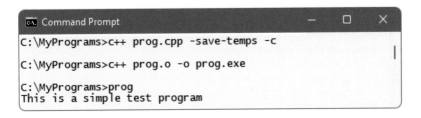

Defining substitutes

Just as the preprocessor substitutes library code for **#include** directives, other preprocessor directives can be used to substitute text or numeric values before assembly and compilation.

The **#define** directive specifies a macro, comprising an identifier name and a string or numeric value, to be substituted by the preprocessor for each occurrence of that macro in the source code.

Like **#include** preprocessor directives, **#define** directives can appear at the start of the source code. As with constant variable names, the macro name traditionally uses uppercase, and defined string values should be enclosed within double quotes. For numeric substitutions in expressions, the macro name should be enclosed in parentheses to ensure correct precedence.

define.cpp

Numeric constants are often best declared as **const** variables – because values substituted by the preprocessor are not subject to type-checking.

1 Start a new program by declaring three **define** directives
```
#define BOOK "C++ Programming in easy steps"
#define NUM 200
#define RULE "********************************"
```

2 Specify the library classes to include, and the namespace
```
#include <iostream>
using namespace std ;
```

3 Add a main function containing a final **return** statement and three statements to output substituted values
```
int main()
{
  cout << RULE << endl << BOOK << endl << RULE ;
  cout << endl << "NUM is: " << NUM << endl ;
  cout << "Double NUM: " << ( ( NUM ) * 2 ) << endl ;
  return 0 ;
}
```

4 Save, compile, and run the program to see the output

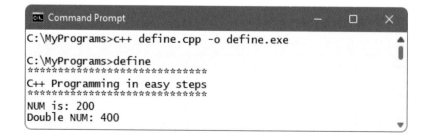

```
C:\MyPrograms>c++ define.cpp -o define.exe

C:\MyPrograms>define
********************************
C++ Programming in easy steps
********************************
NUM is: 200
Double NUM: 400
```

154

5 Recompile the program, saving the temporary files
c++ define.cpp -save-temps -o define.exe

6 Open the temporary "define.ii" file in a plain text editor
and scroll to the end of the file to see the substitutions

```
define.ii - Notepad                          —    □    ×
File  Edit  Format  View  Help
using namespace std ;

int main()
{
  cout   << "*******************************" << endl
         << "C++ Programming in easy steps" << endl
         << "*******************************" ;
         cout << endl << "NUM is: " << 200 << endl ;
         cout << "Double NUM: " << ( (200) * 2 ) << endl;
         return 0 ;
}
```

Substitutions can alternatively be made from the command-line using a **-D***name* option to replace macros with specified values. Note that string values within double-quotes must also be enclosed in escaped quotes in the command – so the substitution will include the double-quote characters.

7 Delete, or comment-out, the **define** directives for both the
BOOK and **NUM** identifiers – then save the program file to
apply the changes

8 Recompile the program, specifying substitute macro
values, then run the program once more
**c++ -DNUM=50 -DBOOK=\""Java in easy steps"\"
define.cpp -o define.exe**

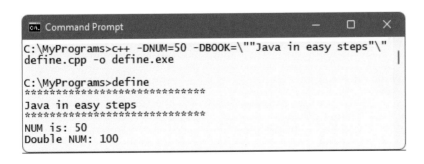

```
Command Prompt                               —    □    ×
C:\MyPrograms>c++ -DNUM=50 -DBOOK=\""Java in easy steps"\"
define.cpp -o define.exe

C:\MyPrograms>define
*******************************
Java in easy steps
*******************************
NUM is: 50
Double NUM: 100
```

Defining conditions

The preprocessor can make intelligent insertions to program source code by using macro functions to perform conditional tests. An **#ifdef** directive performs the most common preprocessor function by testing to see if a specified macro has been defined. When the macro <u>has</u> been defined, so the test returns true, the preprocessor will insert all directives, or statements, on subsequent lines up to a corresponding **#endif** directive.

Conversely, an **#ifndef** directive tests to see if a specified macro <u>has not</u> been defined. When that test returns true, it will insert all directives, or statements, on subsequent lines up to a corresponding **#endif** directive.

To satisfy either conditional test, it should be noted that a **#define** directive need only specify the macro name to define the identifier – it need not specify a value to substitute.

Any previously defined macro can be removed later using the **#undef** directive – so that subsequent **#ifdef** conditional tests fail. The macro can then be redefined by a further **#define** directive:

ifdef.cpp

1 Start a new program with a conditional test to insert a directive when a macro is not already defined
```
#ifndef BOOK
  #define BOOK "C++ Programming in easy steps"
#endif
```

2 Specify the library classes to include, and the namespace
```
#include <iostream>
using namespace std ;
```

3 Add a main function containing a final **return** statement
```
int main()
{
  // Program code goes here.
  return 0 ;
}
```

4 In the main function, add a conditional preprocessor test to insert an output statement when the test succeeds
```
#ifdef BOOK
  cout << BOOK ;
#endif
```

5 Add another conditional preprocessor test to both define a new macro and insert an output statement when the test succeeds

```
#ifndef AUTHOR
  #define AUTHOR "Mike McGrath"
  cout << " by " << AUTHOR << endl ;
#endif
```

6 Next, add a conditional test to undefine a macro if it has already been defined

```
#ifdef BOOK
  #undef BOOK
#endif
```

7 Now, add a conditional test to redefine a macro if it is no longer defined, and to insert an output statement

```
#ifndef BOOK
  #define BOOK "Linux in easy steps"
  cout << BOOK " by " << AUTHOR << endl ;
#endif
```

8 Save, compile, and run the program to see the insertions

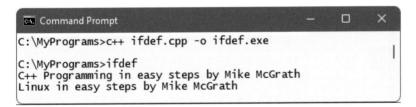

```
Command Prompt                          —    □    ×

C:\MyPrograms>c++ ifdef.cpp -o ifdef.exe

C:\MyPrograms>ifdef
C++ Programming in easy steps by Mike McGrath
Linux in easy steps by Mike McGrath
```

9 Recompile the program, this time defining the **BOOK** macro in the command, then run the program again to see the specified value appear in the first line of output

`c++ -DBOOK=\""Python in easy steps"\" ifdef.cpp -o ifdef.exe`

```
Command Prompt                          —    □    ×

C:\MyPrograms>c++ -DBOOK=\""Python in easy steps"\"
ifdef.cpp -o ifdef.exe

C:\MyPrograms>ifdef
Python in easy steps by Mike McGrath
Linux in easy steps by Mike McGrath

C:\MyPrograms>_
```

Each preprocessor directive must appear on its own line – you cannot put multiple directives on the same line.

On Windows systems, string macro values specified in a command must be enclosed in escaped double quotes.

Providing alternatives

The conditional test performed by **#ifdef** and **#ifndef** can be extended to provide an alternative by adding an **#else** directive. For example:

```
#ifdef WEATHER
  cout << WEATHER ;
#else
  #define WEATHER "Sunny"
#endif
```

Similarly, **#if**, **#else**, and **#elif** macros can perform multiple conditional tests, much like the regular C++ **if** and **else** keywords.

For testing multiple definitions, the **#ifdef** macro can be expressed as **#if defined**, and further tests made by **#elif defined** macros.

While most macros are defined in the source file with a **#define** directive, or on the command line with the **-D** option, some macros are automatically predefined by the compiler. Typically, these have names beginning with a double underscore __ to avoid accidental confusion with chosen names. The compiler's predefined macros are platform-specific, so a program can employ a multiple definition test to identify the host platform:

1. Launch a plain text editor and save a new file (without any content) as "empty.txt" in your program's directory

2. To see a list of the compiler's predefined macros, issue a command calling the cpp preprocessor directly with a "-dM" option on the empty file
cpp -dM empty.txt

3. Scroll through the list to find the "_WIN32" macro on Windows or the "__linux" macro on Linux systems

The **#elif** macro simply combines **else** and **if** to offer an alternative test.

else.cpp

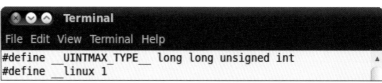

Use the **cpp** command to call the preprocessor directly (not the **c++** command) and ensure the **-dM** option is capitalized correctly.

4 Start a new program with a conditional test to seek the
_WIN32 and __linux macros – to identify the platform

```
#if defined _WIN32
  #define PLATFORM "Windows"
#elif defined __linux
  #define PLATFORM "Linux"
#endif
```

5 Specify the library classes to include, and the namespace

```
#include <iostream>
using namespace std ;
```

6 Now, add a main function containing a final **return**
statement and a statement to identify the host platform

```
int main()
{
  cout << PLATFORM << " System" << endl ;
  return 0 ;
}
```

7 In the main function, insert statements to execute for
specific platforms

```
if ( PLATFORM == "Windows" )
  cout << "Performing Windows-only tasks..." << endl ;
if ( PLATFORM == "Linux" )
  cout << "Performing Linux-only tasks..." << endl ;
```

8 Save, compile, and run the program to see platform-
specific output

The predefined
_WIN32 macro has one
underscore but the
__linux macro has two
underscore characters.

The conditional test of
predefined macros could
be extended to seek
those of other operating
systems, and a final
#else directive added to
specify an "Unknown"
default.

159

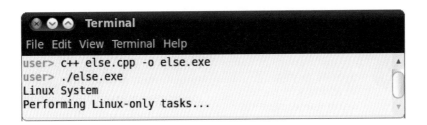

```
Command Prompt                                    —   □   ×
C:\MyPrograms>c++ else.cpp -o else.exe

C:\MyPrograms>else
Windows System
Performing Windows-only tasks...
```

```
⊗ ⊙ ⊙   Terminal
File Edit View Terminal Help
user> c++ else.cpp -o else.exe
user> ./else.exe
Linux System
Performing Linux-only tasks...
```

Guarding inclusions

Typically, a C++ program will have many **.h** header files and a single **.cpp** implementation file containing the main program. Header files may often contain one or more **#include** directives to make other classes or functions available from other header files, and can cause duplication where definitions appear in two files. For example, where a header file includes another header file containing a function definition, the compiler will consider that definition to appear in each file – so compilation will fail.

The popular solution to this problem of re-definition employs preprocessor directives to ensure the compiler will only be exposed to a single definition. These are known as "inclusion guards" and create a unique macro name for each header file. Traditionally, the name is an uppercase version of the file name, with the dot changed to an underscore – for example, **RUM_H** for a file **rum.h**.

In creating a macro to guard against duplication, an **#ifndef** directive first tests to see if the definition has already been made by another header file included in the same program. If the definition already exists, the compiler ignores the duplicate definition, otherwise a **#define** directive will permit the compiler to use the definition in that header file:

Hot tip

Inclusion guards are also known as "macro guards" or simply as "include guards".

160

add.h

triple.h

guard.cpp

1. Create a header file named "add.h" containing the inline declaration of an "add" function
   inline int add (int x, int y) { return (x + y) ; }

2. Now, create a header file named "triple.h" containing a processor directive to make the add function available for use in the inline declaration of a "triple" function
   #include "add.h"

   inline int triple (x) { return add(x, add(x, x)) ; }

3. Start a new program with preprocessor directives to make both the add and triple functions available
   #include "add.h"
   #include "triple.h"

4. Specify the library classes to include, and the namespace
   #include <iostream>
   using namespace std ;

5 Add a main function containing statements that call both the **add()** and **triple()** functions from the included headers

```
int main()
{
  cout << "9 + 3 = " << add( 9, 3 ) << endl ;
  cout << " 9 x 3 = " << triple( 9 ) << endl ;
  return 0 ;
}
```

Beware

Use the conventional naming scheme, where the macro name resembles the file name, to avoid conflicts.

6 Save the files, then attempt to compile the program to see compilation fail because the add function appears to be defined twice – in "add.h" and by inclusion in "triple.h"

```
C:\MyPrograms>c++ guard.cpp -o guard.exe
In file included from triple.h:8,
                 from guard.cpp:6:
add.h:9:14: error: redefinition of 'int add(int, int)'
    9 |     inline int add (int x , int y ) { return ( x + y ) ; }
      |                ^~~
In file included from guard.cpp:5:
add.h:9:14: note: 'int add(int, int)' previously defined here
    9 |     inline int add (int x , int y ) { return ( x + y ) ; }
      |                ^~~
```

7 Edit the header file "add.h" to enclose the inline function declaration within a preprocessor inclusion guard

```
#ifndef ADD_H
#define ADD_H

inline int add ( int x, int y ) { return ( x + y ) ; }

#endif
```

Don't forget

All header files should contain header guards – add a **TRIPLE_H** macro to the **triple.h** file.

8 Save the modified file, then compile and run the program – compilation now succeeds because the inclusion guard prevents the apparent re-definition of the add function

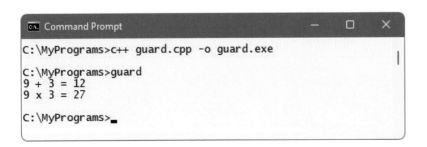

```
C:\MyPrograms>c++ guard.cpp -o guard.exe

C:\MyPrograms>guard
9 + 3 = 12
9 x 3 = 27

C:\MyPrograms>
```

161

Using macro functions

The **#define** directive can be used to create macro functions that will be substituted in the source code before compilation.

A preprocessor function declaration comprises a macro name, immediately followed by parentheses containing the function's arguments – it is important not to leave any space between the name and the parentheses. The declaration is then followed by the function definition within another set of parentheses. For example, a preprocessor macro function to half an argument looks like this:

#define HALF(n) (n / 2)

Care should be taken when using macro functions, because unlike regular C++ functions, they do not perform any kind of type checking – so it's quite easy to create a macro that causes errors. For this reason, inline functions are usually preferable to macro functions, but because macros directly substitute their code, they avoid the overhead of a function call – so the program runs faster. The resulting difference can be seen in the first temporary file created during the compilation process:

macro.cpp

1 Start a new program by defining two macro functions to manipulate a single argument
#define SQUARE(n) (n * n)
#define CUBE(n) (n * n * n)

2 After the macro function definitions, specify the library classes to include, and a namespace prefix to use
#include <iostream>
using namespace std ;

3 Next, declare two inline functions to manipulate a single argument – just like the macro functions defined above
inline int square (int n) { return (n * n) ; }
inline int cube (int n) { return (n * n * n) ; }

4 Add a main function containing a final **return** statement
int main()
{
  // Program code goes here.
  return 0 ;
}

...cont'd

5 At the start of the main function block, declare and initialize an integer variable
`int num = 5 ;`

6 Now, insert statements to call each macro function and each comparable inline function

```
cout << "Macro SQUARE: " << SQUARE( num ) << endl ;
cout << "Inline square: " << square( num ) << endl ;
cout << "Macro CUBE: " << CUBE( num ) << endl ;
cout << "Inline cube: " << cube( num ) << endl ;
```

7 Save the file, then compile the program, saving the temporary files and run the program

`c++ macro.cpp -save-temps -o macro.exe`

```
Command Prompt                              —  □  ✕

C:\MyPrograms>c++ macro.cpp -save-temps -o macro.exe

C:\MyPrograms>macro
Macro SQUARE: 25
Inline square: 25
Macro CUBE: 125
Inline cube: 125
```

8 Open the temporary ".ii" file in a plain text editor like Notepad, to see that the macro functions have been directly substituted in each output statement

```
macro.ii - Notepad                          —  □  ✕
File  Edit  Format  View  Help
using namespace std ;

inline int square( int n ) { return ( n * n ); }
inline int cube( int n ) { return ( n * n * n ); }

int main()
{
  int num = 5 ;

  cout << "Macro SQUARE: " << ( num * num ) << endl ;
  cout << "Inline square: " << square( num ) << endl ;
  cout << "Macro CUBE: " << ( num * num * num ) << endl ;
  cout << "Inline cube: " << cube( num ) << endl ;

  return 0 ;
}
```

Hot tip

Using uppercase for macro names ensures that macro functions will not conflict with regular lowercase function names.

163

Don't forget

An inline function saves the overhead of checking between a function prototype declaration and its definition.

Building strings

The preprocessor **#** operator is known as the "stringizing" operator, as it converts a series of characters passed as a macro argument into a string – adding double quotes to enclose the string.

All whitespace before or after the series of characters passed as a macro argument to the stringizing operator is ignored, and multiple spaces between characters is reduced to just one space.

The stringizing operator is useful to pass string values to a preprocessor **#define** directive without needing to surround each string with double quotes.

A macro definition can combine two terms into a single term using the **##** merging operator. Where the combined term is a variable name, its value is not expanded by the macro – it simply allows the variable name to be substituted by the macro.

strung.cpp

1 Start a new program by defining a macro to create a string from a series of characters passed as its argument, to substitute in an output statement
```
#define LINEOUT( str ) cout << #str << endl
```

2 Define a second macro to combine two terms passed as its arguments into a variable name, to substitute in an output statement
```
#define GLUEOUT( a, b ) cout << a##b << endl
```

3 After the macro definitions, specify the library classes to include, and a namespace prefix to use
```
#include <string>
#include <iostream>
using namespace std ;
```

4 Add a main function containing a final **return** statement
```
int main()
{
  // Program code goes here.
  return 0 ;
}
```

...cont'd

5 At the start of the main function block, declare and initialize a string variable, then append a further string
```
string longerline = "They carried a net " ;
longerline += "and their hearts were set" ;
```

6 Now, add statements to output text using the macros
```
LINEOUT(In a bowl to sea went wise men three) ;
LINEOUT(On a brilliant night in          June) ;
GLUEOUT( longer, line ) ;
LINEOUT(On fishing up the moon.) ;
```

Hot tip

Notice that the second statement contains multiple spaces, which will be removed by the stringizing operator.

7 Save the file, then compile the program, saving the temporary files and run the program
```
c++ strung.cpp -save-temps -o strung.exe
```

```
Command Prompt                           —    □    ×

C:\MyPrograms>c++ strung.cpp -save-temps -o strung.exe

C:\MyPrograms>strung
In a bowl to sea went wise men three
On a brilliant night in June
They carried a net and their hearts were set
On fishing up the moon.
```

8 Open the temporary ".ii" file in a plain text editor, like Notepad, to see that the string values and the variable name have been substituted in the output statements

Don't forget

The merging operator is alternatively known as the "token-pasting" operator, as it pastes two "tokens" together.

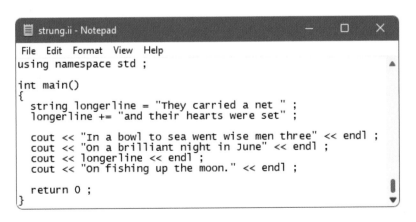

```
strung.ii - Notepad                              —    □    ×

File  Edit  Format  View  Help
using namespace std ;

int main()
{
  string longerline = "They carried a net " ;
  longerline += "and their hearts were set" ;

  cout << "In a bowl to sea went wise men three" << endl ;
  cout << "On a brilliant night in June" << endl ;
  cout << longerline << endl ;
  cout << "On fishing up the moon." << endl ;

  return 0 ;
}
```

assert.cpp

Don't forget

You can comment-out sections of code when debugging using C-style /* */ comment operators.

Beware

Do not place a backslash continuation character on the last line of the definition, and remember to use the # stringizing operator to output the expression as a string.

Debugging assertions

It is sometimes helpful to use preprocessor directives to assist with debugging program code – by defining an **ASSERT** macro function to evaluate a specified condition for a Boolean value.

The condition to be evaluated will be passed from the caller as the **ASSERT** function argument. The function can then execute appropriate statements according to the result of the evaluation. Multiple statements can be included in the macro function definition by adding a backslash \ at the end of each line, allowing the definition to continue on the next line.

Numerous statements calling the **ASSERT** function can be added to the program code to monitor a condition as the program proceeds. For example, to check the value of a variable as it changes.

Usefully, an **ASSERT** function can be controlled by a **DEBUG** macro. This allows debugging to be easily turned on and off simply by changing the value of the **DEBUG** control macro:

1 Start a new program by defining a DEBUG macro with an "on" value of 1 – to control an ASSERT function
#define DEBUG 1

2 Next, add a macro **if-elif** statement block to define the ASSERT function according to the control value
#if(DEBUG == 1)
        // Definition for "on" goes here.
#elif(DEBUG == 0)
        // Definition for "off" goes here.
#endif

3 In the top part of the ASSERT function statement block insert a definition for when the debugging control is set to "on" – to output failure details from predefined macros

```
#define ASSERT( expr )                                  \
cout << #expr << " ..." << num ;                        \
if ( expr != true )                                     \
{                                                       \
  cout << " Fails in file: " << __FILE__ ;              \
  cout << " at line: " << __LINE__ << endl ;            \
}                                                       \
else cout << " Succeeds" << endl ;
```

4 In the bottom part of the ASSERT function statement block, insert a definition for when the debugging control is set to "off" – to simply output the current variable value

```
#define ASSERT( result )                                    \
cout << "Number is " << num << endl ;
```

Predefined macro names are prefixed by a double underscore and suffixed by a double underscore.

5 After the macro definitions, specify the library classes to include, and a namespace prefix to use

```
#include <iostream>
using namespace std ;
```

6 Add a main function containing a final **return** statement

```
int main()
{
    // Program code goes here.
    return 0 ;
}
```

7 At the start of the main function block, declare and initialize an integer variable, then call the macro ASSERT function to check its value as it gets incremented

```
int num = 9 ;    ASSERT( num < 10 ) ;
num++ ;          ASSERT( num < 10 ) ;
```

Additionally, the current date and time can be output from the **__DATE__** and **__TIME__** predefined macros.

8 Save, compile, and run the program to see the output

```
Command Prompt                              —    □    ×

C:\MyPrograms>c++ assert.cpp -o assert.exe

C:\MyPrograms>assert
num < 10 ...9 Succeeds
num < 10 ...10 Fails in file: assert.cpp at line: 30
```

9 Edit the program to turn debugging off by changing the control value, then recompile and re-run the program

```
#define DEBUG 0
```

```
Command Prompt                              —    □    ×

C:\MyPrograms>c++ assert.cpp -o assert.exe

C:\MyPrograms>assert
Number is 9
Number is 10
```

Summary

- The C++ compiler's **-save-temps** option saves the temporary files created during the compilation process for examination.

- Compilation first writes program code and included library code into a single **.ii** text file, which is then translated into low-level Assembly language as a **.s** text file.

- Assembly language **.s** files are translated to machine code as **.o** object files, which are used to create the executable program.

- A **#define** directive defines a macro name and a value that the preprocessor should substitute for that name in program code.

- The preprocessor can be made to perform conditional tests using **#ifdef**, **#ifndef**, and **#endif** directives.

- Preprocessor alternatives can be provided using **#if**, **#else**, and **#elif** directives, and a definition can be removed using **#undef**.

- Each header file should use inclusion guards to prevent accidental multiple definition of the same class or function.

- The macro name of an inclusion guard is an uppercase version of the file name, but with the dot replaced by an underscore.

- A **#define** directive may also define a macro function that will be substituted in program code in place of the macro name.

- Inline functions are usually preferable to macro functions because, unlike macro functions, they perform type checking.

- The preprocessor **#** stringizing operator converts a series of characters passed as a macro argument into a string value.

- Two terms can be combined into a single term by the preprocessor **##** merging operator.

- An **ASSERT** macro function is useful for debugging code, and may be controlled by a **DEBUG** macro to easily turn debugging on or off.

10 Programming visually

This chapter brings together elements from previous chapters to build a complete C++ application in a visual programming environment.

Starting a Universal project

Universal

Windows 10 introduced the **Universal Windows Platform** (UWP) that enables you to create a single **Universal Windows Application** (UWA) that will run on any modern Windows-based device. The interface layout of a UWA uses the **eXtensible Application Markup Language** (XAML) to specify components. In order to develop apps for the UWP, you should be running Windows 10 or Windows 11, and your Visual Studio IDE must include the **Universal Windows App Development Tools**:

1 Go to the Windows apps menu and launch the **Visual Studio Installer**

2 Click the Installer's **Modify** button to open the "Modifying" dialog

Beware

The example in this chapter is described for Visual Studio 2022 on Windows 11 – the instructions may vary for other versions.

3 Select the "Workloads" menu, then choose **Universal Windows Platform development**

4 Click the dialog's **Modify** button to download and install the UWP development workload components

Total space required -3.29 GB

5 Click the **Launch** button to open the Visual Studio IDE

Modify

Launch

Don't forget

Depending upon your choices when you installed Visual Studio, you may already have the Universal Windows App Development Tools.

6 Select the option to **Create a new project**

7 Scroll down the list of project options and click the option to **Install Windows Universal tools for C++ development** to download the tools

8 Select **Create a new project** again and choose the **Blank App (Universal Windows - C++/CX)** project option

Ensure that the Windows **Developer Mode** and **Device discovery** options are enabled in **Settings**, **Privacy & security**, **For developers**.

9 Click **Next** to open a **Configure your new project** dialog, then name the project "Universal" and click **Create**

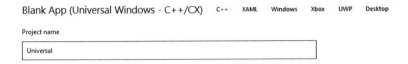

10 When asked to select the target and minimal platform versions, simply click **OK** to accept the default options

11 After Visual Studio creates the new project, select **View**, **Solution Explorer** to examine the generated files:

- A set of logo images in an **Assets** folder.

- Internal XAML and C++ **App** files.

- XAML and C++ files for the **MainPage** – here is where you will create interface components and functional code.

- Other miscellaneous **Package** files.

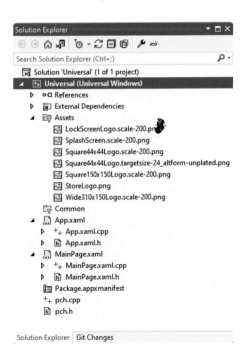

These files are essential to all UWP apps using C++, and exist in every project Visual Studio creates to target the Universal Windows Platform with C++.

Inserting page components

Visual Studio provides a two-part window to insert interface components into a UWP app. This comprises a **Design** view of the components and a **XAML** view for the XAML code:

C++

Universal
(continued)

Hot tip

XAML is pronounced "zammel".

1 Open **Solution Explorer** then double-click on **MainPage.xaml** – to launch the two-part window

2 See that by default, the **Design** view displays a blank canvas in **Landscape** mode

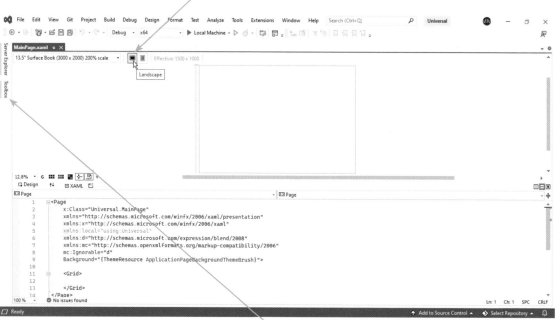

There is a pop-out **Toolbox** feature that lets you add components onto the canvas, but you will need to edit them in the XAML code later. In this example, the components are created in XAML code from the very start.

Now, see that by default, the **XAML** view reveals there are **<Grid> </Grid>** tags – this is the root element of the canvas in which you can add component elements

Component elements are best nested within a **<StackPanel>** element, as this can be given an **x:Name** for reference in functional code and an **Orientation** attribute to specify the direction in which the nested elements should appear. Common component elements include **<Image>**, **<TextBox>**, **<TextBlock>** (label), and **<Button>**. Several **<StackPanel>** elements can be nested within each other to determine the **Horizontal** and **Vertical** layout of components:

 Insert elements between the root **<Grid> </Grid>** tags so the **XAML** view code looks precisely like this:

```xml
<Grid>

    <StackPanel x:Name="MainStack" Orientation="Horizontal">

        <Image x:Name="Image" Width="200" Height="200"/>

        <StackPanel x:Name="Controls" Orientation="Vertical" VerticalAlignment="Center">

            <StackPanel x:Name="Labels" Orientation="Horizontal">
                <TextBlock  x:Name="textBlock1" Text="TextBlock" />
                <TextBlock  x:Name="textBlock2" Text="TextBlock" />
                <TextBlock  x:Name="textBlock3" Text="TextBlock" />
                <TextBlock  x:Name="textBlock4" Text="TextBlock" />
                <TextBlock  x:Name="textBlock5" Text="TextBlock" />
                <TextBlock  x:Name="textBlock6" Text="TextBlock" />
            </StackPanel>

            <StackPanel x:Name="Buttons" Orientation="Horizontal">
                <Button x:Name="BtnPick" Content="Button" />
                <Button x:Name="BtnReset" Content="Button" />
            </StackPanel>

        </StackPanel>

    </StackPanel>

</Grid>
```

The outer **<StackPanel>** is a horizontal layout containing an **<Image>** and a nested **<Stackpanel>**. The nested **<StackPanel>** is a vertical layout containing two further **<StackPanel>** elements that each display their components horizontally. The **x:** prefix before the **Name** attribute refers to the XAML schema used by UWP apps. Notice that each **<TextBlock>** element has a **Text** attribute that can be referenced in functional code. For example, **textBlock1->Text**. references the text contained in the first TextBlock element.

 As you add the component elements in **XAML** view, they appear in the **Design** view until it looks like this:

Importing program assets

In order to have a XAML **<Image>** component display a graphic, an image file first needs to be added to the project's **Assets** folder. It can then be assigned to a **Source** attribute of the **<Image>** tag:

Universal
(continued)

1 Copy an image into the project's Asset folder, typically at
C:\Users*username*\\source\repos*project**project*\Assets

2 Open **Solution Explorer**, then right-click on the **Assets** folder and choose **Add** from the context menu

3 Now, choose **Existing Item...** from the next context menu – to open an **Add Existing Item** dialog box

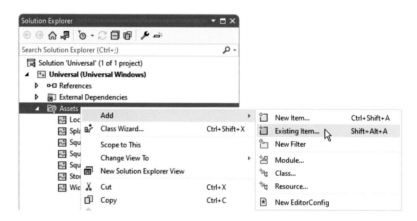

4 In the **Add Existing Item** dialog, browse to the location of the image file and click the **Add** button

An image for display may be in any popular file format – such as **.bmp**, **.gif**, **.jpg**, **.png**, or **.tif**.

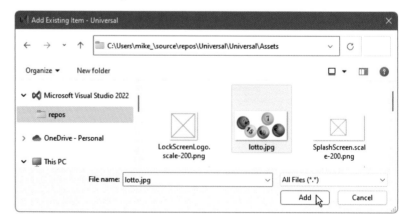

5 In **Solution Explorer**, the selected image file now appears in the project's **Asset** folder

6 Select the Image component in **Designer** view, then click **View**, **Properties** to reveal its properties

7 In the **Properties** window, expand the **Common** category, then click the **Source** item's arrow button and select the added image from the drop-down list

Hot tip

Explore the **Appearance** and **Transform** options in an image's **Properties** window to discover how you can modify how it will be displayed.

8 The image now appears in the **Design** view, and its path gets added to the **XAML** view code and **Source** property

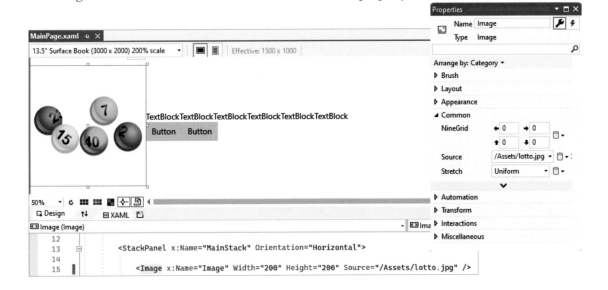

Designing the layout

To complete the app's layout, design attributes can be added to the XAML element tags to specify what they will display and precisely where in the interface they will appear:

 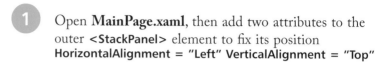

Universal
(continued)

1 Open **MainPage.xaml**, then add two attributes to the outer **<StackPanel>** element to fix its position
HorizontalAlignment = "Left" VerticalAlignment = "Top"

2 Next, edit the **<Image>** element by modifying the initial assigned value of 200 – to increase its width
Width = "300"

3 Now, add an attribute to the nested **<StackPanel>** element to fix its position
VerticalAlignment = "Center"

4 Then, edit all six **<TextBlock>** elements alike, to specify their initial content, width, and margin on all four sides
Text = "..." Width = "20" Margin = "15"
Text = "..." Width = "20" Margin = "15"
Text = "..." Width = "20" Margin = "15"
Text = "..." Width = "20" Margin = "15"
Text = "..." Width = "20" Margin = "15"
Text = "..." Width = "20" Margin = "15"

Hot tip

A single **Margin** value sets all four margins around that component. You can specify two values to set left & right, top & bottom margins; e.g. **Margin = "10,30"**. Alternatively, you can specify four values to set left,top,right,bottom margins individually; e.g. **Margin = "10,30,10,50"**

5 Edit the first **<Button>** element to rename it, specify its button label content, and set its margin on all four sides
x:Name = "BtnPick" Content = "Get My Lucky Numbers"
Margin = "15"

6 Edit the second **<Button>** element to rename it and specify its button label content
x:Name = "BtnReset" Content = "Reset"

7 Finally, add an attribute to each respective **<Button>** element to specify their default state
IsEnabled = "True"
IsEnabled = "True"

...cont'd

The order in which the attributes appear in each element is unimportant, but the elements within the **MainPage.xaml** file should now look similar to the screenshot below:

```xml
<StackPanel x:Name="MainStack" Orientation="Horizontal" HorizontalAlignment="Left" VerticalAlignment="Top">

    <Image x:Name="Image" Width="300" Height="200" Source="/Assets/lotto.jpg" />

    <StackPanel x:Name="Controls" Orientation="Vertical" VerticalAlignment="Center">

        <StackPanel x:Name="Labels" Orientation="Horizontal">
            <TextBlock  x:Name="textBlock1" Text="..." Width="20" Margin="15" />
            <TextBlock  x:Name="textBlock2" Text="..." Width="20" Margin="15" />
            <TextBlock  x:Name="textBlock3" Text="..." Width="20" Margin="15" />
            <TextBlock  x:Name="textBlock4" Text="..." Width="20" Margin="15" />
            <TextBlock  x:Name="textBlock5" Text="..." Width="20" Margin="15" />
            <TextBlock  x:Name="textBlock6" Text="..." Width="20" Margin="15" />
        </StackPanel>

        <StackPanel x:Name="Buttons" Orientation="Horizontal">
            <Button x:Name="BtnPick" Content="Get My Lucky Numbers" Margin="15" IsEnabled="True" />
            <Button x:Name="BtnReset" Content="Reset" IsEnabled="True" />
        </StackPanel>

    </StackPanel>

</StackPanel>
```

As you make changes to the **XAML** view code, the component layout gets changed accordingly in the **Design** view and should now look like this:

Hot tip
You can optionally add **Margin = "0"** attributes to explicitly require elements to have no margin width.

177

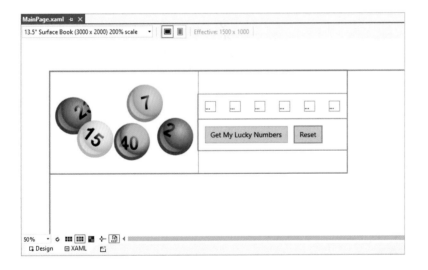

Adding runtime function

Having completed the application component layout with XAML elements on page 177, you are now ready to add functionality with C++ programming code:

Universal
(continued)

1 In **Design** view, double-click on the **BtnPick** button

2 The **MainPage.xaml.cpp** code-behind page opens in the **Code Editor** at a generated **BtnPick_Click** event-handler

3 In the **BtnPick_Click** event-handler block, insert these statements to create a randomized array of integers between 1 and 59

```
int i , j , k , seq[ 60 ] ;
srand( ( int ) time( 0 ) ) ;

for ( i = 1 ; i < 60 ; i++ ) seq[ i ] = i ;
for ( i = 1 ; i < 60 ; i++ )
{
  j = ( ( int ) rand( ) % 59 ) + 1 ;
  k = seq[ i ] ; seq[ i ] = seq[ j ] ; seq[ j ] = k ;
}
// Statements to be inserted here (Steps 4-5).
```

Hot tip

The **srand( )** function seeds a random number generator with the current system time, so subsequent calls to the **rand( )** function return different pseudo-random numbers.

4 Next, insert statements to assign six array element values to the **<TextBlock>** components

```
textBlock1->Text = seq[ 1 ].ToString( ) ;
textBlock2->Text = seq[ 2 ].ToString( ) ;
textBlock3->Text = seq[ 3 ].ToString( ) ;
textBlock4->Text = seq[ 4 ].ToString( ) ;
textBlock5->Text = seq[ 5 ].ToString( ) ;
textBlock6->Text = seq[ 6 ].ToString( ) ;
```

Don't forget

There is no **Label** component in UWP apps; it is called a **TextBlock** instead.

5 Next, insert statements to set the **<Button>** states

```
BtnPick->IsEnabled = false ;
BtnReset->IsEnabled = true ;
```

Beware

There is no **Enabled** property in UWP apps; it is called **IsEnabled** instead.

6 Return to **MainPage.xaml**, then in **Design** view, double-click on the **BtnReset** button

7 The **MainPage.xaml.cpp** code-behind page opens in the **Code Editor** at a generated **BtnReset_Click** event-handler

8 In the **BtnReset_Click** event-handler block, insert statements to assign strings to the **<TextBlock>** components
textBlock1->Text = "..." ;
textBlock2->Text = "..." ;
textBlock3->Text = "..." ;
textBlock4->Text = "..." ;
textBlock5->Text = "..." ;
textBlock6->Text = "..." ;
// Statements to be inserted here (Step 9).

9 Finally, insert statements to set the **<Button>** states
BtnPick->IsEnabled = true ;
BtnReset->IsEnabled = false ;

The **MainPage.xaml.cpp** code-behind page should now look like the screenshot below:

The **BtnReset** button simply returns the **<TextBox>** and **<Button>** components to their original states.

```
MainPage.xaml.cpp  ☐ ✕
Universal                                              ↓ Universal::MainPage
30   □void Universal::MainPage::BtnPick_Click(Platform::Object^ sender, Windows::UI::Xaml::RoutedEventArgs^ e)
31   {
32       int i, j, k, seq[60];
33       srand((int)time(0));
34       for (i = 0; i < 60; i++)seq[i] = i;
35   □   for (i = 0; i < 60; i++)
36       {
37           j = ((int)rand() % 59) + 1;
38           k = seq[i]; seq[i] = seq[j]; seq[j] = k;
39       }
40       textBlock1->Text = seq[1].ToString();
41       textBlock2->Text = seq[2].ToString();
42       textBlock3->Text = seq[3].ToString();
43       textBlock4->Text = seq[4].ToString();
44       textBlock5->Text = seq[5].ToString();
45       textBlock6->Text = seq[6].ToString();
46       BtnPick->IsEnabled = false;
47       BtnReset->IsEnabled = true;|
48   }
49
50   □void Universal::MainPage::BtnReset_Click(Platform::Object^ sender, Windows::UI::Xaml::RoutedEventArgs^ e)
51   {
52       textBlock1->Text = "...";
53       textBlock2->Text = "...";
54       textBlock3->Text = "...";
55       textBlock4->Text = "...";
56       textBlock5->Text = "...";
57       textBlock6->Text = "...";
58       BtnPick->IsEnabled = true;
59       BtnReset->IsEnabled = false;
60   }
```

Notice that the first **for** loop contains only one statement to be executed on each iteration, so braces are not required.

10 Return to the **MainPage.xaml** file, then in **XAML** view, see that attributes have been automatically added to the **<Button>** elements to call the event-handler code

```
<Button x:Name="BtnPick" Content="Get My Lucky Numbers" Margin="15" IsEnabled="True" Click="BtnPick_Click" />
<Button x:Name="BtnReset" Content="Reset" IsEnabled="True" Click="BtnReset_Click" />
```

Testing the program

Having added functionality with C++ code on pages 178-179, you are now ready to test the program:

Universal
(continued)

1 On the Visual Studio standard toolbar, select the **Debug** for **x64** architecture and **Local Machine** options, then click the **Start** button to run the app with debugging enabled

2 Wait while the application gets built and loaded, then click the buttons to try out their functionality

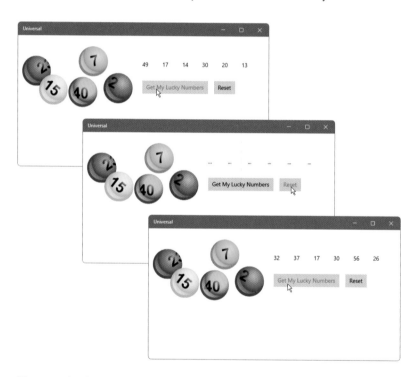

Don't forget

You must have your PC set to **Developer Mode** in **Settings**, **Privacy & security**, **For developers**.

The app looks good – numbers are being randomized and the button states are changing as required.

3 Now, on the Visual Studio standard toolbar, select **Debug**, **Stop Debugging** to exit the running program

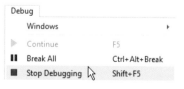

4 Click the **Start** button to restart the app, then click the buttons and compare the sequences of randomized numbers to those generated in the previous test

Great – the random number generator is seeded using the current time to ensure a different sequence each time the app gets run.

5 Drag the window edge to reduce its width, to simulate how the app would look on a narrower display area

This is unsatisfactory – adjustments will be needed to the interface layout so it will adapt to suit narrower display areas.

Adjusting the interface

The app test for narrow display areas on page 181 failed to satisfactorily present the controls, as the interface is too wide for smaller display areas. Happily, the interface can be made to adapt to different sizes so it can also look good on narrow display areas. The adaptation relies upon recognizing the display area size and changing the orientation of a **<StackPanel>** element in **XAML** for narrow display areas:

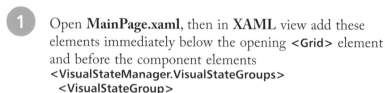

Universal
(continued)

Hot tip

XAML code recognizes the same **<!-- -->** comment tags that are used in HTML code.

Don't forget

Remember that the outer **<StackPanel>** in this app contains an **<Image>** and a nested **<StackPanel>** displayed horizontally, side-by-side. If displayed vertically, they should appear one above the other.

1 Open **MainPage.xaml**, then in **XAML** view add these elements immediately below the opening **<Grid>** element and before the component elements

```
<VisualStateManager.VisualStateGroups>
  <VisualStateGroup>

  <!-- Elements to be inserted here (Steps 2-3) -->

  </VisualStateGroup>
</VisualStateManager.VisualStateGroups>
```

2 Next, insert elements to recognize wide screens

```
<VisualState x:Name = "wideState" >

  <VisualState.StateTriggers>
    <AdaptiveTrigger MinWindowWidth = "641" />
  </VisualState.StateTriggers>

</VisualState>
```

3 Now, insert elements to recognize narrow screens and to change the **Orientation** of the outer **<StackPanel>**

```
<VisualState x:Name = "narrowState" >

  <VisualState.StateTriggers>
    <AdaptiveTrigger MinWindowWidth = "0" />
  </VisualState.StateTriggers>

  <VisualState.Setters>
    <Setter
    Target = "MainStack.Orientation" Value = "Vertical" />
  </VisualState.Setters>

</VisualState>
```

The beginning of the **MainPage.xaml** file should now look similar to the screenshot opposite, top:

...cont'd

```
<Grid>

    <VisualStateManager.VisualStateGroups >
        <VisualStateGroup>
            <VisualState x:Name="wideState">
                <VisualState.StateTriggers>
                    <AdaptiveTrigger MinWindowWidth="641" />
                </VisualState.StateTriggers>
            </VisualState>
            <VisualState x:Name="narrowState">
                <VisualState.StateTriggers>
                    <AdaptiveTrigger MinWindowWidth="0" />
                </VisualState.StateTriggers>
                <VisualState.Setters>
                    <Setter Target="MainStack.Orientation" Value="Vertical" />
                </VisualState.Setters>
            </VisualState>
        </VisualStateGroup>
    </VisualStateManager.VisualStateGroups>
```

4 Select **x64** and **Local Machine** then click **Start** to run the app once more – it still looks and functions well

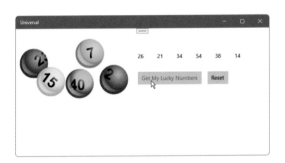

Hot tip

You can have Visual Studio nicely format the XAML code by pressing **Ctrl + K**, **Ctrl + D**.

5 Now, drag the window edge to reduce its width, and see the interface adapt to suit a narrower screen

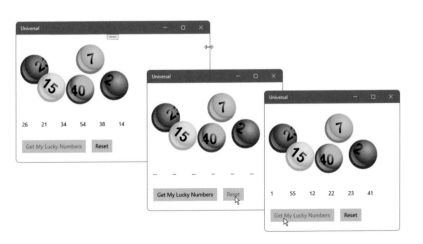

Beware

Although an app may work well in the Visual Studio emulator, it is recommended you always test on an actual device before deployment.

Deploying the application

Having tested the app on page 183, it can now be deployed onto your PC. This will register the app on your PC and add it to the **All apps** menu, so an icon image can usefully be added to the program files to identify the app:

Universal
(continued)

1 Add a 400 x 400 pixel icon image into the project's **Assets** folder on your PC

2 In **Solution Explorer**, right-click on the **Assets** folder and choose **Add, Existing Item** from the context menu and add the icon image

3 Next, double-click **Package.appxmanifest** and select the **Visual Assets tab** and **All Visual Assets** item

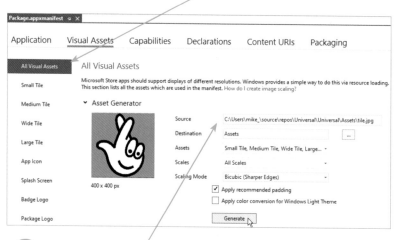

4 Add the icon image as the **Source**, and click the **Generate** button to produce variously-sized tiles and logo images

5 Change the build type to **Release**, then click **Build, Deploy Solution** to register the app on your PC and see it get added to your PC's **All apps** menu

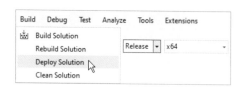

An **App Package** can be created for deployment on other devices running Windows version 1809 or later:

1 Select **Release** on the toolbar

2 In **Solution Explorer**, right-click on the top-level **Universal** project folder, then choose **Publish, Create App Packages** from the context menu to launch a wizard

3 In the "Create App Packages" wizard, choose the **Sideloading** option, then click **Next**

4 Select a package signing certificate, then click **Next**

5 Choose a destination (e.g. a USB drive), version numbering, and architecture – then click **Create**

The build creates a folder of several items, including an Application Package file (**.msixbundle**) and a Security Certificate (**.cer**) file.

6 Move the folder to another device, then right-click on the Security Certificate file and **Install** it in the Trusted Root Certification Authorities certificate store on that device

7 Finally, double-click the Application Package file to **Install** and **Launch** the app on that device

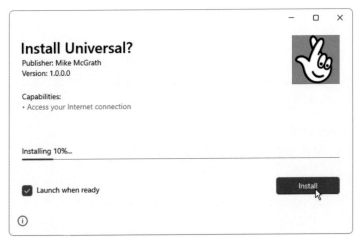

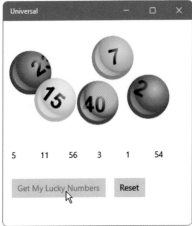

Summary

- The **Universal Windows Platform** (UWP) enables a single app to run on any modern Windows-based device.
- The **eXtensible Application Markup Language** (XAML) is used to specify components and layout on UWP apps.
- The **Universal Windows App Development Tools** are needed in order to develop UWP apps.
- The **Blank App (Universal Windows)** template can be used to create a new UWP project.
- Visual Studio provides a graphical **Design** view and a text code **XAML** view for the **MainPage.xaml** file.
- Component elements can be placed within XAML **<StackPanel>** elements to arrange their orientation.
- Image files can be added to the **Assets** folder and assigned to XAML **<Image>** elements for display on the interface.
- Space can be added around a component by adding a **Margin** attribute and assigned value within its element tag.
- Functional C++ programming code can be added to the **MainPage.xaml.cpp** code-behind page.
- The **Developer Mode** setting must be enabled in the Windows settings in order to develop and test UWP apps.
- A UWP app can be tested in **Debug** mode on the **Local Machine**.
- The interface of a UWP app can adapt to different screen sizes by changing the orientation of **<StackPanel>** elements.
- Image files can be added to the **Assets** folder for assignment as logos in the **Package.appxmanifest** window.
- A **Release** version can be deployed on the Local Machine PC using the **Build, Deploy Solution** menu.
- A **Release** version can be deployed on other modern Windows devices using the **Publish, Create App Packages** wizard.

F

G

H

L